Eilif Dahl had one of the most original and creative minds in plant geography. His approach went far beyond the description of distribution patterns and the establishment of correlations between distributions and particular climatic variables. His understanding of the physiological mechanisms that influenced and controlled the observed distributional patterns was a key feature of his numerous ideas and hypotheses. He was also aware of the importance of history as an influence on present-day plant distributions, especially in arctic plants.

In *The phytogeography of Northern Europe* Dahl brings to bear his wide range of interests in physics, chemistry, geology, climatology, meteorology and mathematics, as well as plant ecology and plant systematics, to analyse and explain the distribution of individual plant taxa across northwestern Europe.

Eilif Dahl died when work on this book was at an advanced stage, and the final version has been assembled by Gro Gulden, his widow and fellow scientist, with support and assistance from John Birks, a long-term friend and colleague. This book will stand as a testament to the ideas and inspiration of a fine scientist.

THE PHYTOGEOGRAPHY OF NORTHERN EUROPE

THE PHYTOGEOGRAPHY OF NORTHERN EUROPE

(BRITISH ISLES, FENNOSCANDIA AND ADJACENT AREAS)

Eilif Dahl

CAMBRIDGE UNIVERSITY PRESS
Cambridge, New York, Melbourne, Madrid, Cape Town, Singapore, São Paulo

Cambridge University Press
The Edinburgh Building, Cambridge CB2 2RU, UK

Published in the United States of America by Cambridge University Press, New York

www.cambridge.org
Information on this title: www.cambridge.org/9780521383585

First published 1998
This digitally printed first paperback version 2007

A catalogue record for this publication is available from the British Library

Library of Congress Cataloguing in Publication data
Dahl, Eilif, 1916–1993
The phytogeography of northern Europe : British Isles,
Fennoscandia and adjacent areas / by Eilif Dahl.
    p.   cm.
Includes bibliographical references (p. mm) and index.
ISBN 0 521 38358 7 (hardback)
1. Phytogeography–Europe, Northern.   2. Plant ecology–Europe Northern.
I. Title.
QK281.D34   1997
581.948–dc21   96-37970   CIP

ISBN-13 978-0-521-38358-5 hardback
ISBN-10 0-521-38358-7 hardback

ISBN-13 978-0-521-03559-0 paperback
ISBN-10 0-521-03559-7 paperback

Contents

Preface

Eilif Dahl had one of the most original and creative minds in plant geography, as well as in plant sociology and mountain ecology. His approach to plant geography went far beyond the description of distribution patterns and the establishment of correlations between distributions and particular climatic variables. He always strove to try to understand what the underlying physiological mechanisms were that influenced and controlled the observed distributional patterns. He proposed that many mountain plants are restricted to high elevations because of their intolerance to high maximum summer temperatures in the lowlands. He suggested that thermophilous species are restricted in their range by temperature-dependent ATP production during dark respiration. He hypothesised that many eastern boreal species are absent from western oceanic areas because of the inability of the species to cope with alternating mild and cold periods in winter and spring. He also strongly championed the importance of history, particularly glacial survival on ice-free refugia (nunataks) of arctic plants during the glacial stages, to explain the present-day distributions of several high-arctic species. Throughout his plant geographical studies Dahl brought to bear his wide range of scientific interests and abilities in, for example, physics, chemistry, geology, climatology, meteorology and mathematics, as well as plant ecology and plant systematics.

In 1986 Eilif Dahl told me he was planning to write a book on the plant geography of Northern Europe. I was naturally excited by the idea of a book by Eilif Dahl that would bring together his life's work and ideas on this fascinating topic. In 1988 the Cambridge University Press issued a contract to him and by 1991 Eilif Dahl was beginning to send me drafts of chapters to read and comment on. In late December 1992 he gave me a near final draft of Chapters 1–10 of the planned 11 chapters. I finished reading and commenting on them in early March 1993 and I returned them to him with comments and suggestions during the week of 8–14 March. Tragically Eilif Dahl died on 17 March 1993, and so he was not able to complete the book in the way he had originally planned. Eilif's widow Gro Gulden has produced the final version of Chapters 1–10 and the associated appendices based on Eilif's manuscripts together with comments and suggestions from several friends and colleagues. Although not containing the final chapter, which was planned to deal with aquatic plants, this volume contains the most important chapters, maps and appendices of the book as originally envisaged by Eilif Dahl. It is presented in this slightly incomplete form because no-one could write the missing chapter with the same degree of scientific originality and breadth as Eilif Dahl. The present book brings together, for the first time, the major ideas and analyses of Eilif Dahl on plant geography of the northwestern European flora including vascular plants, bryophytes and lichens. As Eilif Dahl once wrote in another context in 1985, 'this contribution is made from the basic philosophy that an imperfect work published is more valuable than a perfect work which is not published'.

I conclude this preface by expressing my great debt to the late Eilif Dahl. It was his infectious scientific enthusiasm and challenging ideas that stimulated me to do my PhD on the past and present vegetation of the Isle of Skye between 1966 and 1969. It was his warm friendship and support that helped me feel immediately at home in the Norwegian botanical circle, and it was his series of papers on glacial survival that stimulated me in 1992 to consider statistical approaches to addressing the question: is the hypothesis of survival on glacial nunataks necessary to explain the present-day distributions of Norwegian mountain plants? Naturally Eilif did not agree with the answer I reached, but we both looked forward to several years of friendly discussion about his beloved nunatak hypothesis.

H. J. B. Birks
Bergen, April 1996

Acknowledgements

The Norwegian Research Institute (NRF, formerly NAVF) is thanked for financial support during the writing of this book. Thanks are also due to Nansenfondet for supporting the production of the distribution maps. The background maps were programmed by Michael Angiloff, Norwegian Institute of Land Inventory (NIJOS) based on calculations of temperatures and respiration values for the *Atlas Florae Europaeae* squares accomplished by Dr Oddvar Skre, Norwegian Forest Research Institute (NISK). Berit Hopeland and Åslaug Borgan, Norwegian University of Agriculture (NLH), have drawn the final distribution maps. These maps constitute an important part of the book and the work of this group of persons was very much appreciated by the author. They are all heartily thanked.

Much of the research referred to in the book was performed by former students and colleagues of the author: Dr Yngvar Gauslaa (NLH), the late Professor Olav Gjærevoll (University of Trondheim), Professor Ola Heide (NLH), Kåre Lye (NLH), Per Salvesen (University of Bergen), Dr Oddvar Skre (NISK), and Dr. Stein Sæbø (NLH) in particular. They have in various ways contributed with constructive discussions, valuable assistance, and comments before and during the writing process. Gauslaa, Skre, Gjærevoll, and an anonymous referee have also read and commented on the manuscript at various stages following the death of the author. Their help and support are warmly appreciated.

Some recent information on plant distributions was received from Professor Reidar Elven (University of Oslo) (vascular plants), Professor Kjell Ivar Flatberg (University of Trondheim), Kåre Lye (NLH) (bryophytes), and Dr Einar Timdal (University of Oslo) (lichens). They also read and updated the relevant parts of the manuscript. Flatberg has produced the distribution map of *Sphagnum angermanicum*. I want to express my sincere thanks for this help.

My husband received help and assistance from many persons and institutions during his work with this book. To all those mentioned above and to those not mentioned, because I do not know them, I am sure I can extend Eilif's most heartfelt gratitude.

Personally, I want to thank Professor John Birks, University of Bergen, for having inspired my husband to accumulate important parts of his lifetime's research within 'two covers'. Furthermore he is thanked for critically reading the manuscript at various stages and for giving it the necessary linguistic brush-up. His, and the late Professor Olav Gjærevoll's encouragement and support have been of invaluable help for me in the completion of this book.

Gro Gulden
Ås, May 1996

1 Introduction

Phytogeography, as the name indicates, is the scientific study of the geographical distribution of plants. It is a science combining botany and geography. The subject can be approached in at least two ways.

1 By studying the geographical distribution of assemblages of plants, namely plant communities. This is a **phytosociological** approach.
2 By studying the geographical distribution of individual plant taxa. This is a **phytogeographical** approach in a restricted sense and it is the approach that this book adopts.

Both approaches are relevant and inter-related. Results obtained from phytosociological studies can throw light on phytogeographical problems and vice versa. The methods used are, however, quite different.

The basic material in phytosociological studies is the relevé, i.e. standardised descriptions of vegetation plots with lists of species and information about their relative quantities or abundances. Sets of relevés are combined into vegetation tables which can be analysed by different methods. One method is to attempt a vegetation classification, the approach used in classical plant sociology. Another approach is to consider the vegetation table as an information matrix which can be represented as points in a multidimensional space. By using different methods, invariably employing computers, relations between relevés and species can be studied. This is the ordination or gradient analysis approach (Gauch 1982).

The basic material in the phytogeographical approach in its restricted sense is data on the occurrences of individual plant taxa. These are summarised verbally in floras, indicating where the plant in question might be found. Little information can be presented in this way, however. An alternative approach is to make distribution maps of plant taxa. These can be made in several different ways. Outline maps, where the area of occurrence of the species of interest is shaded, were much used in the early days of phytogeography (Fig. 1).

However, a compiler of a map rarely has sufficient information to state that the species is present in each square kilometre of the shaded area. The

Fig. 1. Distribution of *Rubia peregrina* in Europe in relation to the +4.5 °C isotherm for January. After Salisbury (1926).

map is based on the compiler's interpretation of the available data and invariably contains a certain amount of subjective evaluation.

A more objective way of representing the available information is by entering a dot on the map for each station where there is reliable evidence that the species was found, documented either by a voucher specimen in herbaria, or as literature records by competent botanists. The resulting map is called a **dot-map**. Atlases containing dot-maps of all species within a country or region have now been produced. A pioneer in this type of work was Eric Hultén who produced distribution maps of all vascular plant species in the Nordic countries (Hultén 1950, 1971a). His maps are, however, not pure dot-maps. All stations of rare species or stations in areas where a species is rare or near its limit of range are marked by dots, whereas areas where the species is considered common are shaded. Hultén extended his mapping to the entire Northern Hemisphere for Amphi-Atlantic species (Hultén 1958), for Circumpolar species (Hultén

1962, 1971b), and after his death, maps of the distribution of all other North European species were published (Hultén & Fries 1986).

Large numbers of dot-maps have now been produced. In the British Isles, Perring & Walters (1962) produced dot-maps of all vascular plant species based on a grid of 10 × 10 kilometres. An atlas of distribution of vascular plants in Switzerland was produced by Welten & Sutter (1982), for the former West Germany by Haeupler & Schönfelder (1988), for Belgium by van Rompaey & Delvosalle (1978, 1979), for Denmark by Vestergaard & Hansen (1989), and for the Netherlands by Mennema *et al.* (1985). Dupont (1990) has produced dot-maps of 645 taxa for France on a 20 × 20 km grid scale and Meusel *et al.* (1965, 1978) have provided maps of numerous central European species by combined shading and dot-mapping.

Finally, by cooperation of botanists in all countries in the area covered by *Flora Europaea*, a project to produce dot-maps for all species on a 50 × 50 km square grid is now in progress. So far distribution maps of about 2400 taxa have been published (Jalas & Suominen 1972–94) out of an estimated total of more than 11 000 species.

Dot-maps of individual interesting plant species, or certain phyto-geographical elements have also been produced. These include mosses (Størmer 1969) and lichens (Degelius 1935). An index (*Index Holmiensis*) tries to register all published distribution maps of species.

An enormous amount of data about plant distributions in Europe is thus available. Like any other information it should be used with caution, with due regard to possible errors in compilation and in interpretation. Important is the question about the completeness of information. Some parts of an area concerned may be well explored, resulting in a high density of dots, whereas other parts may be less well explored and have fewer dots. Thus the density of plots might not reflect the real commonness or abundance of the species in question. For example, in *Atlas Florae Europaeae* certain parts of Russia are poorly covered while, for example, Britain and Scandinavia are well covered.

The interpretation of plant distribution patterns

Maps display distribution patterns of the species, and the next question is, how can these patterns be interpreted or explained. Why is *Ilex aquifolium* restricted to the southern and western parts of Europe, whereas *Picea abies* is restricted to more northern and eastern parts of Europe and to higher altitudes in the south?

The explanation may be historical. Perhaps a species is missing in some part of the area because time has not been sufficient for the species to reach the whole area where it could live and compete. But the explanation may also be ecological. The distribution may be limited because the species has a limited ability to adapt to certain climatic or edaphic conditions. For instance, taking the *Atlas Florae Europaeae* map of *Pinus sylvestris* (Fig. 2), Scots pine may be limited in its northern areas because the summers are too cold, in the southwest because the winters are too mild, and in the southeast because the summers are too hot and dry. These are all hypotheses which need closer examination.

In this respect the North European flora has some important features that may be different from the floras of other areas, for example the Mediterranean or the Tropics. It has long been noticed that species density (the number of different species in a given area, often called species diversity or richness) decreases towards the poles. In Svalbard the total number of native vascular plants is 160 (based on Rønning 1964). The number of native vascular plant species and subspecies in Fennoscandia is 1429 (excluding microspecies in the genera *Rubus*, *Taraxacum* and *Hieracium*) (based on Lid 1985). The corresponding number for the

Fig. 2. Distribution of Scots pine (*Pinus sylvestris*) in Europe. After Jalas & Suominen (1972–91).

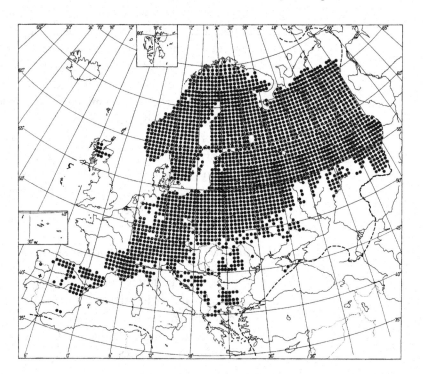

4

British Isles is 1670 (based on Clapham *et al.* 1987). The total number of species treated in *Flora Europaea* is 11 557 and most of these are found in the Mediterranean Region. In comparison, the total flora of Costa Rica in Central America, a country somewhat larger than Denmark, contains 9000 species and important parts are still under-explored. Thus Costa Rica probably has more species than the entire area covered by *Flora Europaea*. The total number of different tree species in the British Isles and Fennoscandia taken together is less than 60. In a park area in Malaysia of approximately 10 hectares, Ashton found 750 different tree species (Wilson 1991). What can the reasons be for the vast differences in diversity between the floras of the Tropics and the flora of Northern Europe? The reason may be ecological, namely that only a few plants are able to adapt to the cool conditions of the north. But more probably, the explanation is historical.

During the last (Weichselian) glaciation, with its maximum ice extent about 18 000 years ago, all higher plants were exterminated in the areas covered by the ice sheet. The vegetation outside the glacial limits was also affected because of the cold and arid climate then prevailing in most of Europe. The lowland flora (with an altitudinal limit below 800 m in southern Norway and below 1800 m in the northern Alps) probably survived the glaciation in the Iberian Peninsula, in the Mediterranean area, and in the areas north and east of the Black Sea and the Caspian Sea and has only partly returned later. The lowland flora element thus has a history of less than 18 000 years in Northern Europe, perhaps even shorter than 12 000 years.

In the Tropics, for example Costa Rica, floral continuity in time goes back to the Cretaceous period. During the glacial periods the climate in areas with tropical rain forests was less affected by the temperature fluctuations associated with the climatic shifts of the Pleistocene than in Northern Europe. The models of climatic conditions during the last glaciation, at the time of maximum glaciation 18 000 years ago, show small temperature differences between present climate and glacial climate for the tropical rain forest areas (Manabe & Hahn 1977; Kutzbach & Guetter 1986), although there may have been important differences in the amount of precipitation. Thus evolution probably has been going on for more than 100 million years under relatively constant climatic conditions in the Tropics, resulting in an amazing diversity (Wilson 1991). There has been time for new taxa to invade already established ecosystems and to adapt genetically to the conditions by coevolution. As suggested by May (1974,

p. 175), the evolution of diversity in an ecosystem takes time. The flora of Northern Europe is young. There has not been sufficient time for any appreciable evolution of new species and subspecies to take place or for invaders to penetrate the existing ecosystems and adapt genetically to the new conditions. The flora is under-saturated in species, especially different tree species.

The problems of biogeography look very different when seen from the perspective of the tropics than from a Northern European perspective. Dynamic biogeography, as defined by Hengeveld (1990), has as its ultimate aim the aiding of our understanding of evolutionary processes. But there has not been much evolution in the Northern European flora during the last 18 000 years. It has not been convincingly shown that any new species has evolved in the area apart from polymorphous apomictic groups or by alloploidy (Dahl 1987, 1989). There is thus not much scope for studying coevolution, a process that may be very important in warmer areas.

The immigration of many species after the last glaciation, especially forest trees, has been reconstructed by palynological investigations (Huntley & Birks 1983; Birks 1986; Huntley 1988) or by combined methods of palynology and macrofossil analysis. About 50% of the present flora of the British Isles has been identified as fossils, and 30% as far back as the late-glacial period (Godwin 1975). Pollen data show that many trees followed close after the retreating ice during deglaciation, whereas others first arrived only when the climate became sufficiently warm. Most of the species appear to have had a sufficient spreading capability to occupy all, or almost all, of the area today where they are able to live and compete (their realised area is thus assumed to be close to their potential area). This suggests that the distribution of species in the lowland flora may be mainly restricted by ecological conditions although the history of this flora must also be considered.

The ecological conditions limiting plant distribution may be **climatic**, depending on meteorological conditions, **edaphic** depending on soil properties, or **biotic** depending upon the associated biota. These are not independent: for example, soil properties are strongly influenced by climate, and the associated biota also have inter-relations with climate. But when considering the distribution of species at a broad continental scale, there is every reason to suppose that these patterns are primarily influenced by climate.

Climatic conditions vary systematically from north to south, from the coast to inland, and from the lowlands to the mountains. Thus the effect of climate on plant distribution can best be investigated by studying distributions at the European scale. Edaphic conditions, depending primarily on the geology, vary at a finer spatial scale. The effect of parent soils on plant distribution can best be studied in limited areas with a varied geology but with a uniform climate. The most difficult task is to evaluate the importance of the biotic component. While climatic conditions and edaphic conditions can be measured, there are at present no adequate means for measuring competition (Grace & Tilman 1990). Competition may be for light, for moisture, or for different critical nutrients, and plants may apply a form of chemical warfare by producing exudates that affect their competitors. The effect of parasites and pathogens may also be important, as well as human interference.

One possible approach to investigate whether the performance of a species is limited by competition is to cultivate a species without competitors and see if it survives. There is no doubt that competition restricts the distribution of many species. The best example is perhaps species which flourished under glacial or late-glacial conditions, but later disappeared when forest invaded. One such species, *Hippophae rhamnoides*, was widespread in Northern Europe (Sandgren 1943; Godwin 1975, p. 207) but is today restricted to coastal areas where wind and salt-spray minimise competition, or to scree communities in the north and in the mountains. Also many weedy species which invade our fields today were widespread and common in the open, late-glacial landscape. They more or less died out in the early Holocene and expanded again with the introduction of agriculture in Neolithic times (Birks 1986).

Competition commonly interacts with climatic factors. With decreasing precipitation forests open up and the leaf area index decreases (Woodward 1987), and as a result low-growing species more resistant to drought can find a niche. It has also been argued that many alpine species are restricted to areas with a cold climate because they cannot compete for light with forest trees in areas of warmer climate. In such cases it is meaningless to ask whether the plants are limited by climate or by competition, because their limitation is truly multi-factorial.

The experimental approach

The problem of interpreting the distributions of plants may be approached in two different ways. One could start in the laboratory, taking a plant species and investigating its physiology and its performance under various combinations of climatic or edaphic factors. Armed with this information one could go into the field and see how far the information helps in explaining plant distribution and abundance (e.g. Forman 1964).

This is a valuable approach, but considerable difficulties are involved. One is that by rigorous testing almost any ecological factor can be found to affect the plants. Moreover the experiment does not tell us which of these factors are really critical for the survival of the plant populations in nature.

Plants respond in different ways to climate and weather. Some limitations are of a kind to which the species cannot adapt with the available gene resources of the population. As an example we may chose the frost-sensitivity limit of a species, for example holly (*Ilex aquifolium*) with a distribution limit corresponding to the −0.5 °C isotherm of the coldest winter month. Holly has grown in Europe since the last glacial age, its diaspores have spread, and wherever it found an unoccupied site with suitable niche conditions, colonisation presumably followed. This went on until the species reached a limit beyond which it could not adapt by selection of biotypes within its gene pool. To try to develop a frost-resistant holly by crossing and selection of European hollies is thus an unpromising enterprise, since nature has performed this experiment on a much larger scale than we will ever be able to.

But plants can adapt to local conditions. For the survival of a population it is necessary that the phenological cycle of the species is synchronised with the annual climatic cycle. It is necessary that the shoots do not break dormancy in spring too early only to be damaged by late frosts, and that the plant flowers, sets seeds, and finally goes into dormancy before the winter. Such phenological cycles are controlled by the responses of the plant to certain seasonal signals, e.g. daylength, temperatures during day and night, etc. If a species has a wide distribution area, local populations from different parts of the area normally have different responses. Local populations form different ecotypes; in forestry they are usually called provenances. By selection of genotypes the populations adapt to local conditions and, given time, can adapt to very different seasonal climates. These responses are considered adaptations to local climate, but only rarely are these factors limiting the distribution of the species.

We have thus to consider two types of ecological factors. The first may be called **primary factors** to which the plants cannot adapt themselves given the available gene resources of the interbreeding populations. The second may be called **secondary factors** to which the populations can adapt, given sufficient time for the development of new ecotypes. The first type of factors must be responsible for the broad features of plant distribution, for example on a country or a continental scale. Good examples are the frost-sensitivity limits. The other type represents factors that are not primarily responsible for the broad-scale distributional limits of the species. Examples of this can be found as differences between ecotypes of a species, as shown by, for example, different edaphic tolerances within a species such as *Hippocrepis comosa*, *Lychnis alpina* and *Minuartia verna*.

In order to measure the physiological potential of a species in the laboratory it is therefore necessary to obtain data for a variety of ecotypes. This complicates the approach of starting in the laboratory in an attempt to interpret the distribution limits of species.

The field approach

The other approach begins in the field. All sites where the plant populations survive are within the ecological niche of the species; the niche is here understood as the set of environmental factors necessary and sufficient for the survival of the population (Hutchinson 1957). Hence the set of observations where the population survives gives us information about the realised extent of the ecological niche. This is usually done by comparing the ranges of the plants with selected environmental parameters to see if there are apparent thresholds beyond which the plant does not grow. One seeks correlations as close as possible between distribution limits and various ecological factors. Sometimes a combination of different factors (indices) is used in such comparisons. The danger of this approach is that by using a sufficiently complicated index one can find a limit that runs along almost any border on the map. Such empirical relations tell us little or nothing about possible cause-and-effect relationships.

By comparing data on species' distribution with environmental measurements correlations can be found. *But a correlation is not a causal relation, it is merely a description of a relation between two variables* (Ratcliffe 1968; Pigott 1975). Such a correlation invites an explanation. If, for example, the northern distribution limit of *Ilex aquifolium* is correlated with a mean temperature of the coldest month of –0.5 °C, it is tempting to interpret this as a cause and effect. This may be possible, but it is not sufficient. One must ask in what way do the cold winters affect holly. If an external factor affects the distribution or performance of a plant, the effect of the external factor must act upon the physiological processes in the plant. *To explain a correlation one also needs to know the physiological mechanisms that are responsible for the correlation.*

The approach used in this book is to start with plant distribution patterns in nature and to look for corresponding patterns in environmental factors that are thought *a priori* to have some important physiological effects on plant growth and survival. From these corresponding patterns hypotheses are generated about the possible physiological mechanisms involved, and these hypotheses are finally tested by physiological measurements and experimental transplants. Pigott (1970) provides a model example of such an approach in his study on *Cirsium acaule* at its northern limit in England.

The following conditions for a satisfactory explanation of a causal relationship between the distribution of a plant, on one hand, and an environmental factor or factors on the other should thus be satisfied (Pigott 1970, 1975):

1 A strong correlation between the distribution and the environmental factor(s) must be present.
2 A physiological mechanism must link the environmental factor(s) to the physiology of the populations of the plant in question.

Additional evidence might be obtained by the study of marginal or outlying populations during years with unusual weather. For example, Iversen (1944) could show that populations of *Ilex aquifolium* in Denmark were virtually wiped out during the cold winters in the 1940s, thereby confirming that winter frost was a limiting factor. But experimental methods can also be used, for example, to transplant populations to experimental gardens outside the natural range of the plant and to observe if they suffer during particular extreme climatic episodes. Dahl (1951) observed that alpine plants planted in the Botanic Garden in Oslo suffered during particularly hot spells, thereby linking poor performance of these plants to high temperatures.

There is a fundamental difference between **production ecology** and **survival ecology**. In ecosystem analysis, one is concerned with the turnover of energy and matter and in applied ecology, as in agriculture, one is interested in maximising the yield. In this type of analysis mean temperatures and duration of the growing season are important parameters, as is the availability of essential nutrients. It is often believed that an efficient production system is also important for the survival of a population, and this may well be true. But it does not help a population when conditions are good during 364 days of a year if the entire population is then wiped out on the 365th. In survival ecology one is often most concerned with extreme events, whereas in production ecology the main interest is often in average conditions.

Climatic correlations

The approach adopted here is primarily correlative. In comparing the broad distribution patterns of the flora with climatic factors, correlations are made with data provided by meteorological stations. Here the measurements are made in standard meteorological screens, usually located 2 m above ground. But plants do not live in meteorological screens, they live exposed to sun and wind. The meteorological data are assumed to be as regionally representative as possible of level ground under average conditions. This is a **macroclimate** or **synoptic climate**. But synoptic climate is modified by topography. Local meteorology, measuring **meso-** or **topoclimate**, is based on observations taken in similar screens as the climatic networks, but the stations for measurements are placed under different conditions of altitude, slope, aspect, etc. But plants grow on the ground and are exposed to radiation. We need to know about the climate within the plants, or in their immediate vicinity, to explore the ecophysiological aspects of climate. This is the field of **micrometeorology**. Here the instrumentation is different: one needs instruments that can measure temperatures in small volumes of air or even within the plant tissue.

In order to elucidate connections between broad climatic features on one hand, and meteorological conditions inside the plants or in their immediate vicinity on the other, models of the heat exchange between plants and the environment are used. Solar radiation is absorbed by the plants. The energy absorbed is dissipated to the environment by the transport of sensible heat to the surrounding air, by evaporation of water, by heating of the soil, and by re-radiation to the environment. These different components can be estimated according to physical laws as a function of meteorological conditions (Dahl 1963a; Gates 1980; Gauslaa 1984) and by means of an energy budget the plant temperature and amount of water evaporated can be calculated. Such calculations show that trees are closely linked to atmospheric conditions. Differences between macro- and microclimate increase with decreasing plant height, especially in open habitats. Thus one would expect closer correlations between the distribution of tree species and macroclimate than between the distribution of plants living close to the ground and the macroclimate.

A local climate in open habitats is generally more continental than the regional climate, with colder nights and warmer days. The largest differences occur on days with little wind and clear skies. Temperature amplitudes are larger in continental areas than in oceanic areas where there is more wind and more overcast cloud. This can create a bias in our evaluations of the correlations between the distribution of low-growing plants and climatic factors.

Floristic elements

When a series of distribution maps are compared, one sees that many species have approximately the same distribution or that the distribution patterns share certain characteristics. For example, Fig. 13 (p. 43) shows the distribution of *Quercus ilex*, Fig. 14 (p. 44) the distribution of *Parietaria judaica*, Fig. 15 (p. 45) the distribution of *Asplenium adiantum-nigrum*, and Fig. 16 (p. 46) the distribution of *Quercus petraea*, in Europe. It will be seen that the areas are broadly similar and that the distribution limits are equiformal progressive. A set of species sharing defined properties, distributional or biological, is called a **floristic element**. A plausible explanation for such distributions could be that all members of the element are limited by a common ecological factor, for example a climatic factor, but that the sensitivity or vulnerability of the species differs. If such grouping can be done without too serious a distortion of the distributional data, an enormous simplification is achieved (e.g. Birks 1987). A general hypothesis can then be made to explain the distribution of many species. In the end the hypothesis must, however, be tested for each species.

The early literature abounds in subdivisions of floras into different distributional elements. The subdivision depends upon the geographical area concerned and what biological or distributional features are considered important. The procedure may be dangerously subjective, and so far only a few attempts have been made to derive or control such subdivisions by numerical ordination or classification methods. Examples of such quantitative attempts in Northern Europe include Proctor (1967), Birks (1976), Pedersen (1990), Myklestad & Birks (1993) and Myklestad (1993).

There is, however, a general agreement about the recognition of at least

four major geographical elements of vascular plants within the North European flora. It should be emphasised that these four major elements form only part of the total North European flora.

1 *An atlantic element.* The taxa of this element have a southern and western distribution in Europe with their optimal representation in the Iberian Peninsula. The distribution limits run from the southeast to the northwest, parallel with the winter temperature isotherms. It is generally agreed that the occurrence of winter frost may be the most important limiting factor.

2 *A warmth-demanding or thermophilic element.* The members of this element have distribution limits in the lowlands running roughly along the geographical parallels from west to east or with a slightly more northern orientation in continental areas in the east than in oceanic areas in the west (e.g. *Quercus robur*, Fig. 33, p. 72). The altitudinal limits of the species are higher in the south than in the north; a good example is the climatic forest limit. It is generally thought that the limiting factor or group of factors is related to the amount of summer heat, and that the species require a sufficiently long and warm summer to ripen their seeds or to ripen the winter buds before the winter frost.

3 *A boreal, northeastern element* with a distribution that is the inverse of the atlantic element. The centre of the species distribution is in northern Russia and Siberia. Typical taxa of the element are those of the boreal conifer forests. Their lower altitudinal limits increase towards the southwest and here the species become subalpine or montane while in the northeast they are abundant in the lowlands. Apparently they require a cold winter to survive. The physiological mechanisms are still obscure, but are probably related in some way to the stability of the winter climate.

4 *An alpine and arctic element* restricted to areas where summers are cool and short. It is commonly assumed that the species are poor competitors that cannot survive in competition with more warmth-demanding lowland species. However, more detailed analysis suggests that other mechanisms may be important. For example, some species are physiologically not adapted to warm conditions and may not survive, even in the absence of competition. The distribution limits are negatively correlated with high maximum summer temperatures (Dahl 1951; Conolly & Dahl 1970).

These four major elements are apparently limited by temperature factors. In addition, there are floristic elements of more or less widespread distribution limited by other factors. Humidity factors may be important for some. There is a **xerophilous element** that is limited to the dry steppe communities in the southern and eastern parts of the North European region. The species in this element have a short growing season in spring or early summer and go into dormancy when the soils become too dry. Many bulbiferous species in deciduous forests also exhibit similar features. Very little is known about their specific correlation with environmental factors. Interactions between soil and climate may be important. At the opposite end, we have the **humidiphilous element**, restricted to areas with moist conditions. An example of this element is *Myrica gale*. Perhaps combinations of requirements for mild winters and summers that are not too warm might explain such distributions. Among the **poikilohydric** bryophytes and lichens many clearly exhibit distribution patterns suggesting that sufficient moisture in the atmosphere is important, for example with occurrences along the coasts in northwest Europe and then further south in the western Alps. The limiting effect of dryness on vascular plants, which are **stenohydric**, is more indirect. As long as there is sufficient water in the soils and evaporation is not too high, water does not limit plant performance.

Finally, there are many rare and widespread plants for which no correlation between distribution and ecological factors is apparent. The distribution for some of them may depend on historical factors as discussed in Chapters 9 and 10, such as presumed glacial survival, human activity, and former land use.

2 Climate

Northern Europe is situated in the low pressure belt between the subtropical high and the polar high. In areas of the Atlantic Ocean that are south and southeast of Iceland the average atmospheric pressure is low and this directs southwesterly winds to Northern Europe. These southwesterly winds bring heat to northern latitudes in western Europe. Warm water transported by the Gulf Stream and North Atlantic Drift also contributes to this. Thus North Europe has, for its latitude, an unusually mild climate. At the same latitude as southern Norway but west of the Atlantic is southern Greenland, where most of the land is covered by an inland ice sheet.

In this climatic system, cyclones are generated that travel eastwards with rain that is enhanced when the cyclones meet the mountains of Scotland, Scandinavia and the Alps. The cyclones have a more northerly position in summer than in winter and the Mediterranean areas thus have drought during summer. In Northern Europe there is cyclonic rain in all the seasons, with a maximum in late summer and autumn. The other major source of precipitation is convective rain, especially during the warm season. On hot days the air near the ground is heated and becomes unstable so that warm air rises up through the atmosphere, is cooled, and liberates precipitation, often as thunderstorms. This is the most important source of rain in the eastern, more continental parts of Europe.

Along the coasts of northwest Europe there is an oceanic temperature regime with a small annual amplitude and rain in all seasons. In areas where winter temperatures are low, much of the precipitation falls as snow which also protects the soil surface from frost penetration in winter, and there is no permafrost. In the interior, the climate is continental with a high annual amplitude, dry summers, and small amounts of snow in winter. In the northeast with cold dry winters, permafrost can develop.

Climatologists describe the climate in terms of parameters which are not always biologically meaningful. Here I shall try to use a set of parameters that are believed to be of biological importance because they are connected with plant physiological mechanisms. The maps are based on primary information contained in Anonymous (1973) and the technical calculations are explained in Appendix I.

The first parameter is **the mean temperature of the coldest month**. This is related to the frost sensitivity of plants and is thus essential to the distribution of atlantic plants. It is also related to the distribution patterns of boreal species. The coldest month in continental areas is January; in oceanic areas near the sea it is February. Figure 3 gives the mean temperature of the coldest month calculated for sea level. It is seen that the warmest winters are in the southwest and the coldest in the northeast. But there are also some effects due to the presence of extensive land areas with large annual amplitudes far from the sea and hence with cold winters.

It might be argued that it is the severity during the coldest days in the year that is most critical to frost-sensitive plants and that minimum temperatures should be used rather than mean temperatures. Minimum temperatures are, however, too strongly influenced by the location of the observational stations, for example whether they are situated in a valley bottom or above the valley. When the Bergen meteorological station was

Fig. 3. Isotherms of the mean monthly temperature of the coldest month (°C) calculated for sea level.

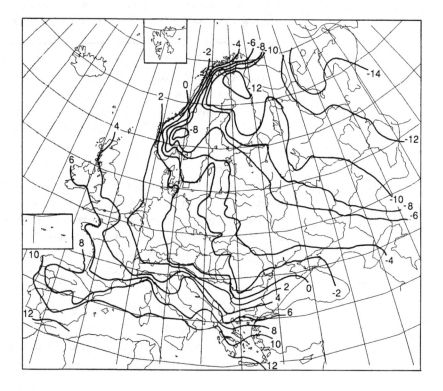

moved from a valley to a hillside the recorded mean minimum temperature increased by 8 °C (Dahl 1951)! Even the mean monthly temperature is not a particularly robust measure. In the valleys of the foothills of the southern Alps and in the Carpathians the mean temperature of the coldest month might be 2 °C below what is observed at a hill station in the same neighbourhood.

The next measure is related to **the amount of heat received in summer**. This is of importance for thermophilic plants. Often the mean temperature of the warmest month is used. Köppen (1920) correlated the timber-line with the 10 °C isotherm for July. Mayr (1909) introduced the mean of the four warmest months (the **tetratherm**) as a better climatic measure. For timber-line observations the mean of the three warmest months (the **tritherm**) seems to be an even better measure (Dahl 1986). Implicit in such measures is that the biological effect of heat has a linear relationship with temperature, while in general, at low temperatures, the relations are curvilinear with an upward curvature.

Fig. 4. Isolines of R-values for the vegetative season calculated for sea level.

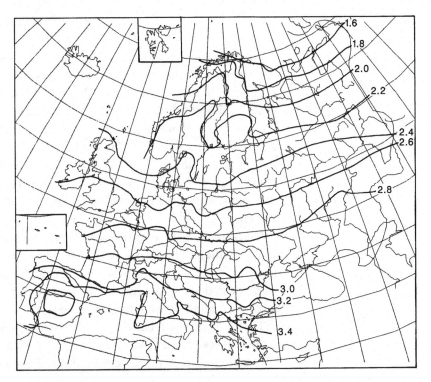

Fig. 5. Isotherms of the mean annual maximum summer temperatures (°C) calculated for sea level.

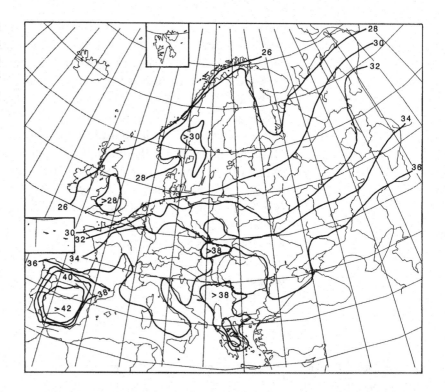

The measures used here for correlations with the distributions of thermophilic plants are related to the respiration hypothesis which is elaborated in more detail later (Chapter 6 and Appendix I). The **respiration equivalent (Re)** is an accumulated sum of temperatures throughout the growing season weighted according to the effect of temperature on dark respiration in plants. The model is based on physiological data for spruce (*Picea abies*). Methods have been developed to calculate such temperature sums and subsequently respiration equivalents from standard data published by the meteorological institutes in different countries (see Appendix I).

I have not used the Re-values on a linear scale, but as natural logarithms (the **R-value**). Figure 4 shows the R-values calculated for sea level. It appears that the isolines run more or less parallel to the meridians but with some exceptions. In general, higher summer temperatures in continental areas are balanced by a longer growing season in coastal areas.

The third measure I shall consider is based on the **mean maximum annual temperatures** recorded at meteorological stations and is relevant to the distribution of arctic-alpine species. Figure 5 gives this measure calculated

Fig. 6. Isolines of the total mean precipitation in the growing season (defined as those months with a mean temperature above +5 °C).

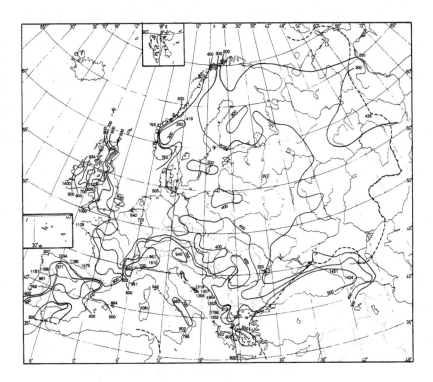

for sea level. This shows lower temperatures in the north and particularly high values in southeast Russia and over Spain and the Balkans.

These three different measures all relate to the temperature component of climate. But humidity factors are also important. Often maps of annual precipitation are published, but precipitation in winter, when plants are dormant, is of little biological importance. Figure 6 presents the total mean precipitation in Europe in the growing season.

Ability to resist drought stress is important for the survival of plants, and to this end they have different strategies. Some survive dry periods as seeds or bulbils. Others are able to conserve water by reducing transpiration. This, however, requires mechanisms to take up the carbon dioxide produced by respiration, so that the pH of the cells remains stable; this involves the formation of lactones from organic acids. Only a few groups, for example succulents, are able to use this strategy. Others have a very deep root system utilising the remaining water resources in the soil (**euxerophytes** *sensu* Iversen 1936). Such plants often have vertical leaves or reflective hairs that reduce the absorption of solar energy and a well-developed sclerenchyma preventing the wilting of the leaves and shoots

Fig. 7. Isolines of the drought index (potential evaporation – precipitation) in the driest month (mm H_2O) calculated for lowland stations.

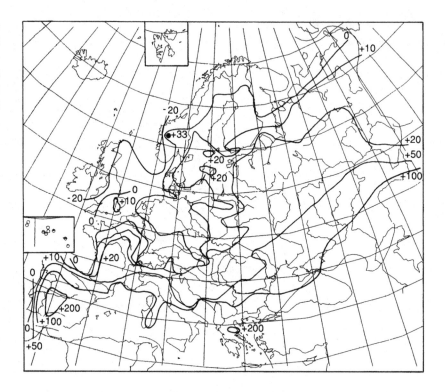

when turgor is reduced. Typically they have thin twigs or small and finely divided leaves and are thus efficient heat exchangers. Solar radiation absorbed by such plants is dissipated to the environment as sensible heat and does not need the expenditure of water to keep temperatures down (Dahl 1966).

Drought affects the competitive structure of some plant communities. When water, temperatures and nutrients are supplied in optimal quantities, competition is for light. This means that species forming a high canopy, such as trees or climbers, absorb the available light, leaving little room for smaller plants. But when moisture becomes limiting, plants respond by reducing the leaf area. Woodward (1987) showed that the leaf area index (LAI = relation between leaf area and ground area) can be predicted from hydrological data. With low amounts of water available to the vegetation, the leaf area becomes smaller, and more light reaches the ground, thus providing suitable niches for smaller plants.

It is during dry spells in summer that plants wilt and die from drought. They are unable to resist a drought stress. As a measure of drought stress, the difference between potential evaporation and precipitation (E – P) can be used.

Figure 7 gives this measure calculated for the driest month at lowland stations in average years. In the north and west precipitation exceeds potential evaporation. Here early summer is the driest season. In the south and east, potential evaporation exceeds precipitation, especially in the interior of Spain and Greece. Here August tends to be the driest month. In Britain there is an area with significant drought stress in the southeast. In Scandinavia this occurs along the eastern coast of southern Sweden, and in some more local areas in the inner fjords and inland valleys of Norway.

Vegetation does not suffer drought during average months, but during particularly dry summers. However, the average values presented in Fig. 7 are believed to be a reasonable reflection of the relative drought stresses within Europe.

3 Edaphic factors

According to Jenny (1941) the soil at any given locality is a product of parent materials, climate, biota, aspect and time. One should perhaps add still another factor, input from the air, such as sea spray bringing in magnesium and sodium, or particles from desert areas such as the Sahara bringing in calcium. Today we must also add pollutants from human activity as an important pedogenic factor.

In the soil, mineral particles release plant nutrients by weathering. Different primary minerals contain different ions and some minerals are more resistant to weathering than others. By far the most important ion in northern European soils is calcium. It is the major ion that neutralises organic acids produced by the breakdown of litter, and the availability of calcium controls the pH of the soil layers.

The most easily weatherable minerals are calcite and dolomite. Of the silicates, plagioclases, especially calcium-rich plagioclases, some amphiboles, and epidote release calcium relatively easily on weathering. Less weatherable are the pyroxenes, and still less weatherable are the potassium feldspars and quartz. In general, parent materials from basic rocks (that is with a low silicon content) such as amphibolite and gabbro provide better soils than parent material from acid rocks such as granite, acid gneiss and sandstone. Of the micas, biotite provides potassium relatively easily by weathering, whereas muscovite is more resistant. The rate of weathering also depends upon the surface areas of the mineral particles. A fine-grained parent material, for example a clay, provides more nutrients to plants than a coarse-grained sand under otherwise equal conditions.

Many plant species are restricted to areas with calcite-bearing parent soils; such species are termed **calcicolous** or **calciphiles**. Others can grow on very poor soils derived from sandstone or acid gneiss and granite; such species are termed **oligotrophic**. Some even avoid soils with a high amount of calcium, such are called **calcifuges** or **calciphobes**. Many species are unable to grow on the poorest soils, but they do not require calcite-bearing parent soils; they form a category between calcicolous and oligotrophic species. Such species have been called, for the lack of a better term, **eutrophic** (Dahl 1957) or 'species that avoid the poorest soils' (McVean &

Ratcliffe 1962). Ellenberg's '*Stickstoffzahl*' (Ellenberg *et al.* 1991) is probably an expression of this gradient.

The microbiological processes are different at high and at low pH. For instance, autotrophic nitrifying bacteria and the most efficient nitrogen-fixing bacteria seem to be unable to function at low pH. Thus the nitrogen cycle in the ecosystem is different in acid and in more neutral soils (Tamm 1991). The amount of carbon in relation to nitrogen in the humus, the C/N ratio, generally decreases with increasing base saturation. In base-rich soils, free nitrate is frequently found in the cell sap of the plants. Eutrophic and calcicolous species normally have a mesomorphic structure, whereas oligotrophic plants generally display a xeromorphic morphology and anatomy, with small and stiff leaves that do not easily wilt, with a high amount of supporting tissue, and with a thick cuticle (Dahl 1957). Such plants are found in heath communities, and also in mire communities, where water is always available and the xeromorphy can hardly be explained as an adaptation to drought. The hypothesis has been proposed that this is an adaptation to a lack of nutrients, primarily of nitrogen but also of phosphorus, a hypothesis reviewed by Sæbø (1970) and Kinzel (1982).

There are also other mechanisms where calcium is apparently essential to plants, and these problems have been reviewed by Kinzel (1982) and Rengel (1992). Uptake of ions by the roots will only function with calcium in the soil solution, but in a balanced nutrient medium plants can grow with amounts of calcium far lower than ever occur in nature. In the soil the situation is apparently different. The uptake of many ions by the roots is against an activity gradient, hence uptake requires energy, and the only energy source for this is ATP supplied by mitochondrial respiration. It is possible that Ca-calmodulin is involved, and that there is a competition between calcium and aluminium ions for absorption sites. Runge & Rode (1991) found that oligotrophic species exhibited a different reaction pattern in relation to the Ca/Al ratio in a nutrient solution compared with more eutrophic species. However, it is still not possible, from physiological measurements in a laboratory, to predict the degree of dependence that a species has for calcium.

At high pH iron and manganese may become unavailable due to their low solubilities. A high demand for iron and manganese might be a reason why calciphobous plants are unable to grow on base-rich soils. This is shown by cultivation experiments where, for instance, *Calluna vulgaris* can be grown on a high pH medium provided that iron is present in a chelated form that can be utilised as a source of iron. Gries (1991) has shown that grasses growing on soils where iron is not easily available excrete chelating substances from their roots, thereby facilitating the uptake of iron.

Climate also affects soils. Throughout most of Northern Europe there is an excess of precipitation over evaporation (Fig. 7). Hence water is transported from the atmosphere through the soil to the ground water. Rain water dissolves ions which are transported out of the ecosystem. The typical soil profile is a podsol with an acid humus on top, a leached layer below, a B-horizon with brown or dark colours due to the accumulation of iron, aluminium and carbon, and a lower layer that is not visibly differentiated. Ions are brought back to the surface through uptake by the plant roots. However, with high excess precipitation, there is always a net loss of nutrients. In general podsolised soils have a low nutrient capital.

If evaporation becomes equal to precipitation, the circulation is reversed and ions are transported upwards. Thus in areas with high drought stress, calcium is conserved in the soils and there may even be an accumulation of soluble salts, for example magnesium sulphate. Such soils are found locally in the driest valleys in southern Norway, where some lichens occur with a highly disjunct distribution (Kleiven 1959). In areas between the dry and the wet, brown soils (brunisols according to the Canadian System: Canadian Soil Survey Committee 1978) are climax soils.

In the FAO–UNESCO soil system, Fennoscandia is dominated by podsols, with lithosols in the mountains and histosols in areas of impeded drainage. In the southern part, on areas below the marine uplift limit, cambisols occur. Within the British Isles, eastern Scotland is an area of podsols, whereas histosols dominate in the west. In northern England and in the northern part of Ireland gleysols represent the typical profile while cambisols dominate in Wales and the rest of England (FAO–UNESCO 1981).

Within each major soil region there is local variation. Pearsall (1950) introduced the term 'flushing' to cover all processes that counteract podsolisation. It could be dry flushing where fresh mineral particles are brought in, for example, as slopewash down hillsides or by rivers in flow depositing minerals on the ground along the banks. It could also be wet flushing where ground water comes to the surface. This may be permanent around springs, or intermittent during periods of high precipitation or during snowmelt where the drainage capacity of the underlying soils is insufficient. Ground water with dissolved plant nutrients thus comes to the surface and replenishes the upper horizons in the soil profile with ions. Later, in summer the soils are again drained. This is typical at lower levels of long slopes, often resulting in tall-herb meadows with a typical brunisol.

Besides calcium, other ions can influence plant distribution. Outcrops of magnesium-rich rocks such as serpentines have a specialised flora, as do heavy-metal areas. Nitrogen is an important factor and many plants are restricted to areas with an ample supply of nitrogen. These are called nitrophilous plants. The available amounts of phosphorus may also limit the performance of many plant species. It may sometimes be difficult to decide whether nitrogen or phosphorus is the limiting factor. On farms which have been abandoned, weeds like *Urtica dioica* and *Chenopodium bonus-henricus* persist for long periods. If nitrogen is added to an ecosystem, for instance by fertilising, the effect lasts only a short time before the added nitrogen is incorporated into the nitrogen capital of the ecosystem. The effect of phosphorus persists much longer and this suggests that phosphorus may be the limiting component. Naturally, along the seashores a halophilous flora is found.

Ellenberg *et al.* (1991) and Landolt (1977) have provided index numbers for different plant species to express their dependence on different soil factors, along with a light index number to express their tolerance to shade, a temperature number to express the position of the species in relation to an altitudinal and/or latitudinal temperature gradient, a continentality number to express the position of the species on a continentality gradient, especially in relation to winter temperature, a moisture number to express the position of the species in relation to moisture, a reaction number to express the position of the species in relation to pH in the soil, a nitrogen number to express the position of the species in relation to the supply of nitrogen, and a salt number to express the relation of the species to sodium. There are no numbers to express the relation of the species to phosphorus or potassium.

The supply of ions for plant growth can vary enormously within a local area. By mapping at the scale of *Atlas Florae Europaeae* with units of 50×50 km, there is usually considerable variation in edaphic conditions within each square. Therefore it is difficult to detect any influence of edaphic factors on the broad-scale distribution patterns, at least within an area like Northern Europe. The effect of edaphic factors on the distribution and abundance of plants can better be analysed by intensive fine-scale mapping within limited areas with a relatively uniform climate.

4   The geological history of the present European flora

The late Tertiary

Here a short account will be given of the development of the flora during the late Tertiary period and the environment within which this development took place. This theme has been elaborated in many text-books and reviews, for example West (1977), Hantke (1978), Nilsson (1982), Birks (1986), Watts (1988) and Tallis (1991).

During the **Miocene** epoch about 15 million years ago the flora of the Northern Hemisphere including Europe contained a mixture of tropical, subtropical and temperate elements. The tropical elements are now extinct in Europe. Of the subtropical and temperate taxa many are now extinct in Europe, but still some persist in North America, the Caucasus, and eastern Asia, for example *Taxodium*, *Sequoia*, *Tsuga*, *Pseudotsuga*, *Glyptostrobus*, *Sciadopitys*, *Ginkgo*, *Liriodendron*, *Liquidambar*, *Nyssa*, *Pterocarya*, *Zelkova* and *Eucommia*.

During the following epoch, the **Pliocene**, the climate gradually became colder. The reason for this might have been the upheaval of mountain massifs in central Asia and western North America affecting the planetary waves, thereby creating conditions for the accumulation of ice and the development of glaciers in the Northern Hemisphere (Ruddiman & Raymo 1988). During this period the tropical elements died out in Europe while several of the subtropical North American and East Asiatic exotics survived. *Tsuga*, *Magnolia*, *Carya*, *Pterocarya*, *Eucommia*, *Parrotia* and *Paeonia* even survived the earliest glaciations of the following epoch, the **Pleistocene**.

During the middle and late Pliocene lowlands existed in the present North Sea, probably extending westwards to Scotland, the Faeroes, Iceland and Greenland and northwards to the Vøring Plateau southwest of Lofoten, North Norway (Rokoengen & Rønningsland 1983; Eldholm & Thiede 1986a, b). In the early Pleistocene, the North Sea was formed by subsidence of the floor along fault lines at about 600 m depth in the south and 1000 m in the north. This picture is supported by studies of molluscs in the Polar Basin and the North Atlantic (Strauch 1970, 1983). The molluscan faunas in the two basins were separated until the isolation was broken in the Pleistocene. Migration between the basins began during the time of the deposition of the Red Crag in East Anglia about 1 million years ago in the early Pleistocene. A barrier for marine molluscs must perforce also have been a bridge for land plants.

Fossiliferous deposits from northernmost Greenland dated to the transition between the Pliocene and the Pleistocene (2 million years ago) contain a flora of American subarctic tundra trees (*Larix* sp. related to *L. occidentalis, Picea mariana, Thuja occidentalis, Cornus stolonifera, Myrica gale* and *Taxus* sp.) together with a subarctic–low-arctic flora with, inter alia, *Betula nana, Dryas octopetala, Vaccinium uliginosum* ssp. *microphyllum, Ledum decumbens,* and a scapiflorous *Papaver* (see Funder *et al.* 1985, Bennike 1987, and Funder 1989 for lists of fossil taxa). The Arctic Ocean at this time lacked a perennial ice cover.

The Pleistocene

The cooling trend continued during the Pleistocene which saw long, cold glacial stages alternating with warm interglacials. During the glacial stages large ice sheets covered much of Northern Europe. Besides the many earlier studies of geological sections on land, important evidence for the changes over time has recently come from studies of drilling cores from the deep ocean. Here sedimentation is believed to have been continuous over long periods and fossil organisms, especially foraminifera, give valuable information about environmental conditions.

The amount of the oxygen isotope ^{18}O in benthic foraminifera is a good measure of the amount of water stored in continental glaciers. From such data it is seen that the interglacials, with a climate warmer or similar to present-day climate and with temperate forest covering the North European lowlands, were of relatively short duration, a few tens of thousands of years. Most of the time the climate was colder than today and **interstadials** (warmer periods with withdrawal of the ice) with forests of *Betula, Pinus, Picea* and *Larix* gave way to **full-glacial stages** when unglaciated areas in Northern Europe were treeless. Maximum glaciation tends to come at the end of the glacial periods (Birks 1986).

The shifts between cold glacial periods and warm interglacials seem to be related to the Milankovitch orbital cycles. During the early Pleistocene, there was a periodicity of 41 000 years, whereas during the last four or five glacial periods there was a periodicity of 100 000 years (Ruddiman & Raymo 1988).

With the advent of the first glacial stages in the Pleistocene, the flora in the interglacials became dominated by temperate genera such as *Quercus, Ulmus, Carpinus* and *Alnus,* and also with more cold-tolerant genera as *Betula, Picea, Pinus* and *Abies.* In the following Cromerian interglacial *Tsuga, Azolla* and *Pterocarya* survived, in the Holsteinian interglacial *Azolla, Osmunda cinnamomea, Rhododendron ponticum* (in Ireland) and *Pterocarya,* and in the Eemian interglacial *Thuja occidentalis, Brasenia purpurea* and *Dulichium arundinaceum* survived. A species of *Picea,* with pollen morphology close to modern *P. omorica,* and *Bruckenthalia spiculifolia* persisted into the early interstadials of the Weichselian glacial stage.

The maximum Pleistocene glaciation

Where ice covered the ground no highly organised plants or animals survived. Within such areas today, all species are immigrants. It is therefore of some interest to map the maximum extent of the ice sheets during the Pleistocene.

The ice limit in the lowlands of Europe is relatively well known from geological studies. In some areas the Saale glaciation was the most extensive, in other areas the Elster glaciation. Figure 8 shows the extreme limit of the Pleistocene glaciation in the west European lowlands and the extreme limit of the Weichselian glaciation.

A question of some importance to phytogeography is whether unglaciated enclaves existed where plants and animals could survive, either as nunataks (see p. 111) protruding through the ice sheets or as unglaciated coastal areas. In general, geologists have been reluctant to accept that such refuges existed along the shores of the North Atlantic Ocean during the last or previous glaciations. Using stratigraphical methods they have been unable to locate such unglaciated areas, and therefore believe that they did not exist. For example, Denton & Hughes (1981) map a complete glaciation of all areas north of southern England and Ireland during the last glaciation, the Weichselian.

However, other methods have recently been developed which permit a mapping of the areas that remained ice-free during the Pleistocene glacial maximum. A feature that frequently can be observed on high mountains in Scandinavia, Britain and elsewhere is **autochthonous boulder fields** ('*Felsenmeere*'). These are accumulations of boulders weathered out from the underlying parent rocks. The usual explanation is that they have been formed by rapid frost-splitting during post-glacial time. However, there is

28

Fig. 8. Extreme limit of Pleistocene glaciation in the west European lowlands (thick line), extreme limit of the Weichselian glaciation (thin line), the limit of the Younger Dryas (Ra) stage in Fennoscandia and the contemporaneous Scottish Loch Lomond Readvance (broken line). Stars and shading show areas with nunataks during the maximum pleistocene glaciation.

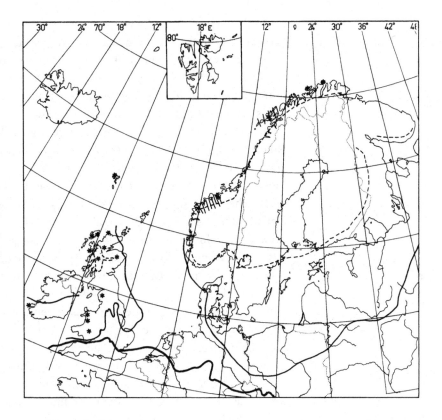

no independent evidence that frost-splitting is especially intense today in areas with such boulder fields.

These boulder fields are restricted to high altitudes in the inland mountains in Scandinavia, about 1700 m in the Jotunheimen mountains. Moving northwest towards the coast such boulder fields can be found at much lower altitudes and the character of the deposits also changes. The boulder fields grade into deposits which in Britain have been called **mountain-top detritus**. It is a rather coarse-grained but not necessarily bouldery autochthonous deposit. The boulder fields with associated mountain-top detritus form a relatively easily mappable unit on the geological maps for western Norway (Sollid *et al.* 1980; Longva *et al.* 1983; Nesje *et al.* 1987, 1988, 1994; Larsen *et al.* 1988).

Occurrences of autochthonous mountain-top detritus in areas with gneissic parent rocks are associated with a deep chemical weathering of gneissic rocks, as shown by analysis of clay minerals. The typical minerals resulting from weathering have been shown to be fine-grained smectite, probably formed from feldspar or other minerals, and vermiculite or hydrobiotite formed by the leaching of biotite. Kaolin and illite are generally absent. This represents a weathering far more advanced than that associated with post-glacial weathering (Dahl 1954, 1961). The fine-grained mountain-top detritus contains the same minerals, suggesting that it has been derived from the weathered rocks.

Of particular importance is the presence of gibbsite at Stadt (Dahl 1961; Roaldset *et al.* 1982; Longva *et al.* 1983) and at Hustad in western Norway (Dahl 1987), on Hadseløy in North Norway (Dahl 1992a), in the mountain-top detritus at Cairngorm, Scotland (Mellor & Wilson 1989) and on the Isle of Skye (Ballantyne 1994), and in saprolites in southeastern Canada and New England (LaSalle *et al.* 1985). Gibbsite is a lateritic mineral that forms in warm climates. This suggests that mountain-top detritus may be a remnant of the Tertiary weathering crust, and the following explanation can be proposed for the origin of the autochthonous boulder fields.

Before the Pleistocene the land was covered with a deeply weathered crust formed during the Tertiary. When the climate became colder, solifluction began. In areas with the coldest climate, solifluction removed the fine-grained fractions of the weathering crust, but the coarser core-stones remained in place. In areas where solifluction was less intense, as in areas near the coast, some of the more fine-grained materials were left in place. The boulder fields are not a result of frost-splitting.

Based on such evidence I have attempted a reconstruction in Fig. 8 of the areas that were unglaciated during the maximum Pleistocene glaciation. In Britain there were a few coastal enclaves in Scotland, for example Tolsta Head in Lewis (Outer Hebrides) and Cape Wrath in Sutherland. Some of the higher mountains of Scotland, for example in the northwest Highlands (McCarroll *et al.* 1995), the summit of Ben Nevis, the Cairngorms, and Ben Wyvis, and on the Inner and Outer Hebrides (Ballantyne 1990; Ballantyne & McCarroll 1995), and in the English Lake District, for example High Street, protruded as nunataks. Considerable parts of the Welsh Mountains were probably nunataks, as were parts of McGillicuddy's Reeks and the Dingle Peninsula in southwest Ireland. Dahl (1992a) has given an isohypse map of the reconstructed surface of the inland ice during the maximum Pleistocene glaciation in northwestern South Norway. In North Norway the Lofoten and Vesterålen islands were areas not reached by the inland ice, but had partial local glaciation only. Also the northernmost headlands of the Faeroes seem to have escaped glaciation.

The Weichselian glacial stage

The limit of the Weichsel inland ice during its maximum, about 18 000 radiocarbon[1] years ago, in Britain and the lowlands of Northern Europe is given in Fig. 8. Some areas in northern Scotland (Caithness and Aberdeenshire) as well as the Orkneys were probably unglaciated, while an ice-cap existed in Shetland (Sutherland 1984; Bowen & Sykes 1988). In Ireland considerable areas remained ice-free. The limit in western and northern Norway is unknown although it is likely that some areas that were glaciated during the maximum Pleistocene glaciation remained ice-free during the maximum Weichselian glaciation. At Andøya in North Norway sections have been found with plant fossils up to 20 000 radiocarbon years old (Vorren *et al.* 1988; Alm & Birks 1991). The fossils suggest a vegetation perhaps similar to that found today in Svalbard and northern Novaya Zemlya.

In the east, between the White Sea and the Urals, the Syrzi moraines are correlated with the Brandenburg moraines in Germany (Andersen 1981), which again are considered to represent the maximum Weichselian glaciation.

Along the coast of West Spitsbergen (Svalbard) shore terraces have been found which could be dated by the amino-acid racemisation method to several hundred thousand years old. Since the terraces are found *in situ* they could not have been overridden by a later inland ice (Miller 1982; Forman & Miller 1984). Thus potential refuges for plants must have existed over several glacial ages and possibly during the entire Pleistocene epoch.

Plant remains in deposits dating back to Weichselian full-glacial time tell us something about the flora in the unglaciated areas south of the North European ice sheets. Apparently the land was unforested between the Alps and the north. The remains indicate dry conditions with a steppe vegetation and a rich fauna of ungulates, for example horses. According to Frenzel (1960, 1968, 1987) and Birks (1986) the climate may be compared with that of the dry steppes of Mongolia today where a combination of low temperatures and drought results in treelessness. It is not to be compared

---

[1] Radiocarbon dates are, by international convention, expressed as radiocarbon years (^{14}C) years before present (BP) where the datum point is AD 1950. Because of past fluctuations in the ^{14}C/^{12}C ratio in the atmosphere there is not a simple linear relationship between radiocarbon years and calendar years. It is now possible to calibrate radiocarbon dates and express them as calendar years for the last 9000 years. Such calibrations rely on radiocarbon dating of tree rings whose calendar age is known.

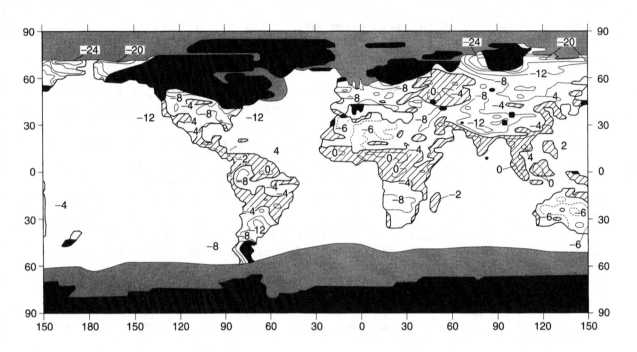

Fig. 9. The reconstructed difference in August mean temperatures between conditions during the Weichselian maximum glaciation and present-day conditions. After Manabe & Hahn (1977).

with the present-day tundra areas north of the arctic timber-line. In the Netherlands and southern England halophilous species occurred far from the coast, indicating a dry climate (Florschutz 1958). The extensive loess sheets formed during the glacial stages could only be formed in a very dry climate. The mire plants present were those of calcicolous mires, and the more acidophilous *Sphagna* occurred more sparsely (Rybníček 1973). There are no indications of heath communities in the full-glacial.

A numerical reconstruction of the August climate 18 000 years ago has been provided by Manabe & Hahn (1977), using as boundary conditions the topography (including the topography of the ice sheets), albedo, and surface temperatures of the oceans provided by a group of geologists and geophysicists (the CLIMAP group). Figure 9 shows the difference in modelled temperature from present-day conditions. The data suggest that summers in most of Europe were 8 °C cooler than at present, and that conditions were dry over much of Europe during the glacial maximum. A reconstruction by Kutzbach & Guetter (1986) that also takes into account differences in solar radiation gives essentially the same result for the region.

Conolly (in Conolly & Dahl 1970) found that all occurrences of subfossil arctic and alpine plants in Britain could be explained by the assumption that the maximum summer temperatures were depressed by at least 6 °C. Analysis of the distribution of *Salix herbacea* in Europe gives the same result (Dahl 1987). It is possible that the discrepancy between the estimates given by Manabe & Hahn (1977) and the fossil data can be explained by the fact that the CLIMAP group assumed the existence of a glacier more than 1000 m thick covering the North Sea, which is contradicted by later information. Kutzbach & Guetter (1986) used the reconstruction of Denton & Hughes (1981) for the extent of the ice sheet 18 000 years ago which is even more extensive than that of the CLIMAP reconstruction. In the Mediterranean area the temperature reconstruction suggests that summers were 7 °C cooler than today, hence the vegetation zones were depressed more than 1000 m. This is confirmed by a study of endemism in the Mediterranean (Dahl 1987).

The climatic depression in terms of winter temperatures may have been greater than for summer temperatures, with temperatures down to –25 °C in northwest Europe (Birks 1986, p. 15). A model reconstruction by Kutzbach & Guetter (1986) of January temperatures 18 000 years ago suggests temperature differences from the present in the order of 20 °C. This fits well with observations by Kolstrup (1980) on fossil ice-wedges in the Netherlands, northern Germany and Denmark, and similar observations by West (1977) from southern England, as mean surface temperatures well below 0 °C are required to develop permafrost. This furthermore fits with the presence of boreal plant species outside their present range (*Betula nana, Pinus sylvestris*), suggesting a 10 °C depression of winter temperatures near Biarritz, southwest France (see p. 91). Reconstructions from Byelorussia (Soffer 1990) suggest very cold winters, with January temperatures of –30 °C and only a short growing season, but with July temperatures not much different from today.

The late-glacial period

The Weichselian glacial stage ended with the **late-glacial** period about 13 000–10 000 radiocarbon years ago. The model reconstructions by Kutzbach & Guetter (1986) suggest a rapid rise in both summer and winter temperatures after maximum glaciation, to temperatures higher than the present day around 9000 years ago. After a cold period, the Older Dryas period, there was a warmer interval, the Allerød, during which glaciers retreated from south Sweden at a rate of 60 m/year (Berglund 1979). This was followed by a colder period, the Younger Dryas period. During the Younger Dryas period the glaciers advanced, both in Britain (the Loch

Lomond re-advance) and in Scandinavia. The extent of the glaciers of the Younger Dryas period is given in Fig. 8. The snow-line in southwest Norway was 400–550 m lower than today (Andersen 1954), whereas in the Gaick mountains in northeast Scotland the snow line was 740–815 m lower (Sissons 1974). Evidence from fossil alpine plants in Britain suggests that summer temperatures were at least 3 °C lower than today (Conolly in Conolly & Dahl 1970). A similar estimate is also obtained from southern Scandinavia (Dahl 1964).

The immigration of the flora after the Weichselian glacial stage

The **Holocene**, or the post-glacial epoch, began after the late-glacial 10 000 radiocarbon years ago. The climate rapidly became warmer and the ice front retreated 300–500 m/year from north of the Middle Swedish moraines (Berglund 1979; Strömberg 1985) and in southeast Norway about 120–150 m/year (Sørensen 1982). The mountains emerged and the last remnants of ice were left in the valley bottoms, damming up lakes ('ice-dammed lakes'). Temperatures rose quite suddenly and the flora closely followed; already by 8400 radiocarbon years BP (before present, taken as 1950) large pines were found in the valleys of eastern Norway, suggesting summer temperatures at least as high as today (Sandmo 1960; Birks 1990).

During the Weichselian glacial maximum no highly organised plants could survive in areas covered by the ice. Moreover, outside the glaciated areas the flora was affected by the severe climatic conditions. With summer temperatures about 6 °C cooler than today, the altitudinal vegetation zones in Scandinavia, the Alps and the Mediterranean were depressed more than 1000 m. This suggests that species which today, for physiological reasons, cannot grow at high altitudes (above 1000 m in southern Scandinavia or 1500 m in the Alps) had no ecological niche in Europe north of the Alps. It must have been too cold for them. The lowland species must have immigrated from refuges in the Mediterranean area including Spain (Huntley & Birks 1983; Birks 1986; Bennett *et al.* 1991). The more acidophilous and oceanic components may have survived along the Atlantic seaboard in Ireland and western France. Species intolerant of high temperatures may have survived in refuges in the British Isles and along the coast of Norway. Others, especially the boreal elements, probably survived east and southeast of the North European ice sheet in Russia.

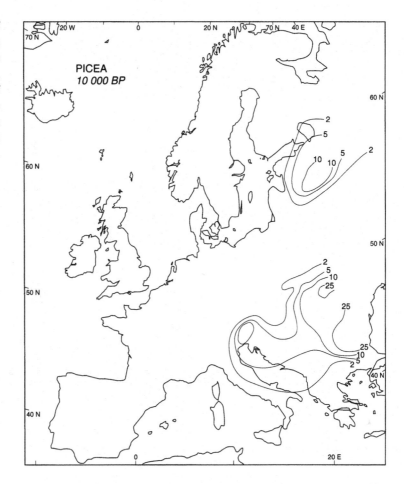

Fig. 10. Isopoll maps of *Picea abies* in Europe. The maps show the pollen percentages of *P. abies* in deposits of different ages before present (radiocarbon years). After Huntley & Birks (1983).

The immigration of elements that can be identified as pollen can be followed by palynological methods, and the immigration of the flora has been beautifully mapped by Huntley & Birks (1983), Peterson *et al.* (1979) and Huntley (1988). Figure 10 (pp. 35–7) gives a series of isopoll maps of *Picea abies* in Europe (from Huntley & Birks 1983). The maps show the pollen percentage of *Picea abies* in deposits of different ages before present. In late-glacial times, *Picea* was present in the Alps and also in Siberia. Following climatic warming the areas expanded, within two separate areas in the Alps and in northeastern Europe. Finally, about 3000 radiocarbon years BP the two areas began to merge, and today the distribution is continuous.

Expansions of the different taxa that can be identified by their pollen can be mapped, their approximate migration rates can be estimated, and their

Fig. 10. (*cont.*)

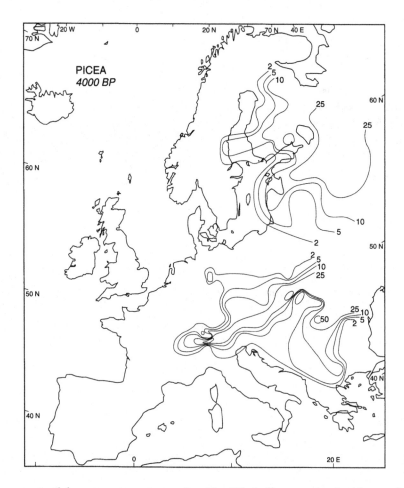

potential **source areas** can be identified. Temperate deciduous forest species of the following genera probably came from the Mediterranean area: *Corylus, Quercus, Fraxinus, Acer, Tilia, Alnus glutinosa* and *Ulmus*. *Abies, Picea* and *Ulmus* may have spread from the Alps, and *Picea, Alnus* cf. *incana, Ulmus, Larix* and *Hippophae* may have come frome refuges in Russia. Probably the tree birches, *Juniperus* and *Populus* remained in the lowlands of Northern Europe. There are some indications that Scots pine (*Pinus sylvestris*) could have survived in western Scotland (Huntley & Birks 1983, p. 628; Kinloch *et al.* 1986; Birks 1989).

The immigration depended partly on the ability of the species to migrate and reach sites where they could grow and reproduce, and partly on environmental conditions (Birks 1986). Species with light, wind-dispersed seeds such as *Betula* and *Pinus* expanded rapidly following the climatic

Fig. 10. (*cont.*)

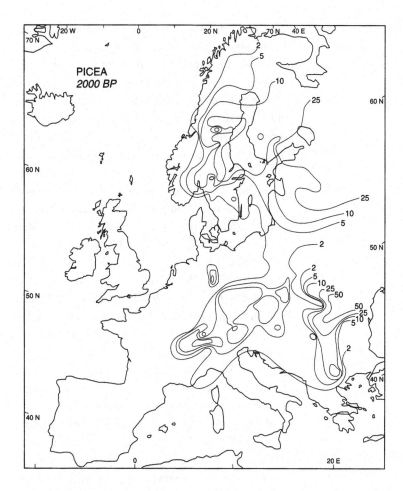

amelioration. Thermophilic, eutrophic, aquatic and marsh taxa also spread early.

The first immigrants occupying the deglaciated areas are termed pioneers and the initial immigration phase is the so-called **protocratic phase** (Iversen 1958; Birks 1986). The vegetation was open with soils of a high pH. There were alpine-subalpine species, indicating open conditions and a cool climate. Species of the dry steppes of the periglacial areas belonging to *Artemisia*, Chenopodiaceae and *Helianthemum* were also present, as were present-day weed species such as *Centaurea cyanus* and *Plantago lanceolata*. A typical pioneer species was *Hippophae rhamnoides* together with several other species with nitrogen-fixing root symbionts. During the warmer Allerød period *Betula pubescens* coll., *Populus tremula* and *Juniperus communis* were present in England and southern Scandinavia, whereas *Pinus sylvestris* had probably reached Bornholm. All these species have diaspores which are easily spread by animals or by wind.

As climate improved and forest trees were able to invade and develop a closed forest canopy the pioneer species disappeared. Only in places where the vegetation remained open, for example in karst areas, could the full-glacial and the late-glacial pioneer flora survive. This was the case on islands in the Baltic, for example Öland and Gotland (Iversen 1954), and in Teesdale in northern England (Godwin 1975) where a relict flora still exists with several endemic and/or disjunct taxa.

During the next immigration phase, the **mesocratic phase**, the climate became warmer than present-day climate. This phase lasted from 9000 to about 2500 radiocarbon years BP and is divided into the Boreal (9000–7000 radiocarbon years BP), the Atlantic (7000–5000 radiocarbon years BP), and the Subboreal (5000–2500 radiocarbon years BP) periods. The lowlands of Northern Europe were colonised by mixed deciduous forests of *Corylus*, *Quercus*, *Ulmus* and *Tilia*. The first to come was *Corylus*, later followed by *Quercus*, *Ulmus* and *Tilia*, and usually still later *Alnus glutinosa*. The warmest period may have been the Atlantic. Forest trees such as *Betula pubescens* and *Pinus sylvestris* grew at altitudes 250–300 m higher in Scandinavia than today (Smith 1920; Kullman 1988; Birks 1990; Faarlund & Aas 1991), and 200–300 m higher in the Alps (Huntley 1988). In the same period many thermophilic trees like *Corylus*, *Quercus*, *Tilia* and *Alnus glutinosa* extended further north than today. The occurrence of fossil pollen of *Viscum album* outside its present geographical range indicates a warmer climate, the difference from present-day summer temperatures might have been around 2 °C (see Fig. 32, p. 71). The winters were also milder. The presence of pollen of *Ilex aquifolium*, *Hedera helix* and *Viscum* about 100 km north of Oslo (Hafsten 1956) indicates winter temperatures perhaps 6 °C warmer than today. *Picea abies* is a boreal species avoiding areas with January temperatures higher than –2 °C. The broad corridor between the areas in the Alps and farther north at 6000 radiocarbon years BP (see Fig. 10) might be explained by winters there being about 4 °C warmer than today (see discussion, p. 91). During the mesocratic phase agriculture and animal husbandry came to the area and introduced new weeds.

Around 4000 radiocarbon years BP a climatic shift took place with the onset of cooler and moister conditions, which culminated in a maximum of the mountain glaciers in parts of Europe around 1750, the so-called 'Little Ice Age'. Since then temperatures have been gradually increasing.

In this phase, termed the **telocratic** (Iversen 1958; Birks 1986), an expansion of acidophilous vegetation types, an expansion of mires, and the immigration of *Picea abies* and *Fagus sylvatica* took place. *Picea abies* expanded gradually from Finland but was never able to colonise Scania in southernmost Sweden and Denmark. It came very close to its present limit in central Norway about 2000 radiocarbon years BP. Later it came to the lowlands of southeast Norway and it may still be slowly extending its range (Fægri 1950; Hafsten 1991). In this period *Fagus sylvatica* reached southern Scandinavia and England, but it has hardly reached its climatic limit since it grows well when planted and is able to reproduce by seeds in western Norway and in Scotland (Birks 1989).

5   The atlantic and oceanic elements

The atlantic and oceanic elements consist of species with a southern and western distribution in Europe. The definitions of these elements have been thoroughly discussed by, for example, Troll (1925), Holmboe (1925), Kotilainen (1933), Degelius (1935), Fægri (1960), Ratcliffe (1968) and Størmer (1969).

In dealing with the atlantic and oceanic elements it is important to distinguish between **stenohydric** and **poikilohydric** plants. Almost all vascular plants are stenohydric; if their cells dry out they die. Lichens and bryophytes are poikilohydric, as their cells quite normally dry out without causing serious damage. The ecology and distribution patterns of stenohydric and poikilohydric plants are quite different and to emphasise these differences, I will use the term **atlantic** for stenohydric plants and **oceanic** for the poikilohydric plants.

The atlantic element

Climatic correlations

An isotherm map of the temperature in the coldest month, calculated for the lowest point in each square grid in *Atlas Florae Europaeae*, is given in Fig. 11. The distribution patterns of atlantic plants are correlated with these isotherms.

Within the atlantic element a number of sub-elements and groups can be recognised according to the correlations between the distribution limits of the species and the winter temperature isotherms. The following sub-elements can be recognised (see Appendix II which lists all the correlations).

1 The British–Mediterranean sub-element

This consists of species limited in Britain and Ireland to areas with winter temperatures higher than +6 °C. *Isoetes histrix* (Fig. 12) is an example.

Fig. 11. Isotherms for the coldest month of the year (°C) calculated for the lowest points in the landscape.

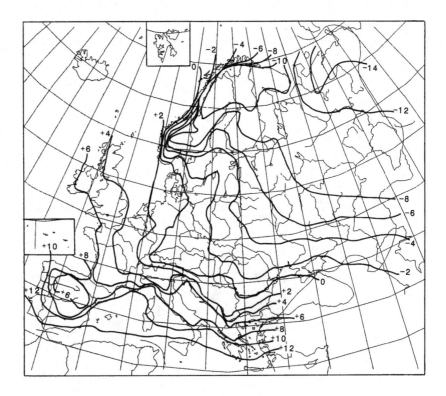

Within this sub-element there are two different groups.

(a) *A British–Channel Islands group* of species restricted to the Channel Islands, the western part of the British south coast, and occasionally occurring on the southernmost coasts of Wales and Ireland. This element is strongly dominated by small annuals (e.g. *Isoetes histrix, Trifolium molinieri, T. bocconei, T. occidentale, Ornithopus perpusillus, Bupleurum baldense, Echium plantagineum*), several annual grasses, and also some geophytes such as *Romulea columnae, Gladiolus illyricus* and *Orchis laxiflora*. Their habitat is typically dry cliff ledges, areas with shallow soils or sand-dunes that dry out in summer, thereby preventing competition from more aggressive species. This is a typical Mediterranean element depending on a combination of mild winters and dry summers.

(b) *A Hibernian group* of species best represented in western Ireland. This consists of acid soil or forest species which are hemicryptophytes or chamaephytes. Many occur along the west coast of France and represent the northernmost occurrences of species widely distributed

41

Fig. 12. Distribution of *Isoetes histrix* in relation to the +6 °C isotherm for the coldest month of the year. Filled circles show species presence, open circles the other grid squares of *Altas Florae Europaeae* where the coldest month at the lowest point has +6 °C or more. Points show presence in grid squares outside the indicated isotherm.

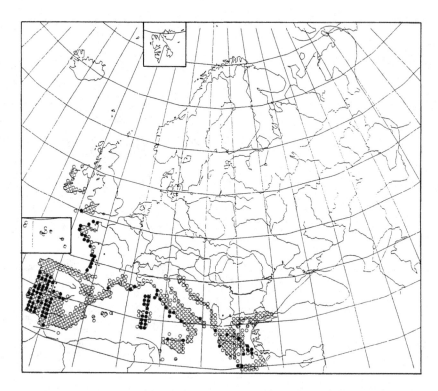

in the Mediterranean. Examples include *Arbutus unedo, Daboecia cantabrica, Erica erigena, E. ciliaris, E. vagans* and *Neotinea intacta*. They may have immigrated during a period with milder winters during the post-glacial; *Erica erigena* may have been a human introduction (Foss & Doyle 1990). Some other species called *Lusitanian* are missing in France but appear in the Iberian Peninsula, for example *Saxifraga spathularis, S. hirsuta* and *Erica mackaiana. Minuartia recurva* and *Pinguicula grandiflora* also show a disjunction to the western Alps. They have an altitudinal limit of several hundred metres in Ireland, an altitude not expected for species limited to areas with high winter temperatures. It has been speculated that they might have survived the Weichselian glaciation somewhere in Ireland, perhaps somewhere along the coast of Connemara (Mitchell & Watts 1970). Fossiliferous deposits from the Gortian (= Holsteinian) interglacial contain fossils of *Daboecia cantabrica, Erica ciliaris, E. mackaiana* and *Rhododendron ponticum* (Coxon & Waldren 1995). Late Pleistocene and Holocene fossils of *Arbutus unedo* have been found in

Eire (Mitchell 1993) and northern France (Meusel *et al.* 1978). The species of the American element in the British Isles, including *Eriocaulon septangulare*, *Sisyrinchium bermudianum* and *Spiranthes romanzoffiana*, are suspected glacial survivors.

2 The British–Atlantic sub-element

The species in this sub-element are restricted to Britain and both sides of the British Channel but do not extend eastwards as far as Denmark and western Norway. They are limited by the winter isotherms of +2 °C or warmer. Examples are *Quercus ilex* (as a naturalised tree) (Fig. 13) and *Parietaria judaica* (Fig. 14). This sub-element consists of many forest or forest margin species such as *Tamus communis* and *Hyacinthoides non-scripta*, heath-land species such as *Ulex europaeus*, and some species of mires or other wet habitats such as *Hypericum elodes*, *Anagallis tenella* and *Wahlenbergia hederacea*.

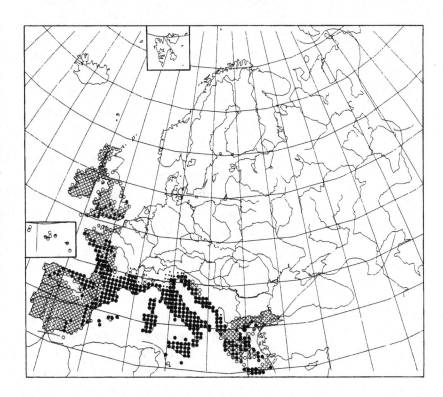

Fig. 13. Distribution of *Quercus ilex* in relation to the +4 °C isotherm for the coldest month of the year. Naturalised occurrences are marked as crosses. Filled circles show species presence, open circles the other grid squares of *Altas Florae Europaeae* where the coldest month at the lowest point has +4 °C or more. Points show presence in grid squares outside the indicated isotherm.

Fig. 14. Distribution of *Parietaria judaica* in relation to the +2 °C isotherm for the coldest month of the year. Filled circles show species presence, open circles the other grid squares of *Altas Florae Europaeae* where the coldest month at the lowest point has +2 °C or more. Points show presence in grid squares outside the indicated isotherm.

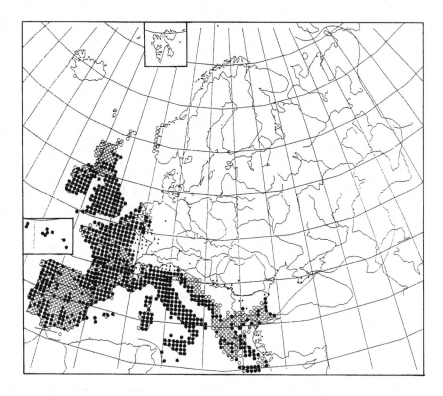

3 The West Scandinavian–Atlantic sub-element

The species in this sub-element are limited by isotherms of –2 °C or warmer. An example is the distribution of *Asplenium adiantum-nigrum* (Fig. 15). In Norway this sub-element has been called hyperatlantic with species like *Asplenium marinum, Dryopteris pseudomas, Luzula sylvatica, Vicia orobus, Ilex aquifolium* and *Erica cinerea.* The species extend as far east as the coast of southwest Sweden, Denmark and Germany.

4 The Scandinavian–Atlantic sub-element

The species in this sub-element are limited by an isotherm of –4 °C or warmer. They grow along the coast in East Germany to about Gdynia. An example is *Quercus petraea* (Fig. 16). Other examples include *Fagus sylvatica, Hedera helix, Erica tetralix* and *Narthecium ossifragum.* They have been called subatlantic species, together with the next sub-element.

Fig. 15. Distribution of *Asplenium adiantum-nigrum* in relation to the −2 °C isotherm for the coldest month of the year. Filled circles show species presence, open circles the other grid squares of *Altas Florae Europaeae* where the coldest month at the lowest point has −2 °C or more. Points show presence in grid squares outside the indicated isotherm.

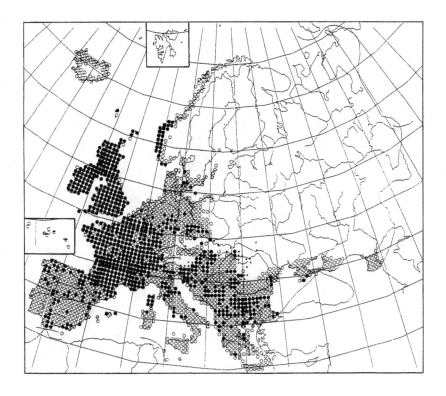

5 The Baltic–Atlantic sub-element

This sub-element is limited by the −8 °C isotherm or warmer and contains species that occur in the Baltic countries and adjacent Russia. Examples are *Cerastium semidecandrum* (Fig. 17), *Blechnum spicant*, *Reynoutria sachalinensis*, *Allium ursinum*, *Dentaria bulbifera* and, as an extreme, *Viscum album*, limited by −8 °C (Fig. 32, p. 71).

Ecophysiology

The underlying physiological mechanisms for the kind of distribution exhibited by the species of the Atlantic element are reasonably clear. It seems that winter frost is limiting although this needs to be demonstrated in individual cases. The problems of frost resistance and its measurement have been summarised by Sakai & Larcher (1987). The sensitivity to frost can be measured by exposing plants or plant parts to defined levels of cold and studying the survival of the different organs. Thus winter buds may have a different cold resistance than the cambium. The resistance depends

Fig. 16. Distribution of *Quercus petraea* in relation to the –4 °C isotherm for the coldest month of the year. Filled circles show species presence, open circles the other grid squares of *Altas Florae Europaeae* where the coldest month at the lowest point has –4 °C or more. Points show presence in grid squares outside the indicated isotherm.

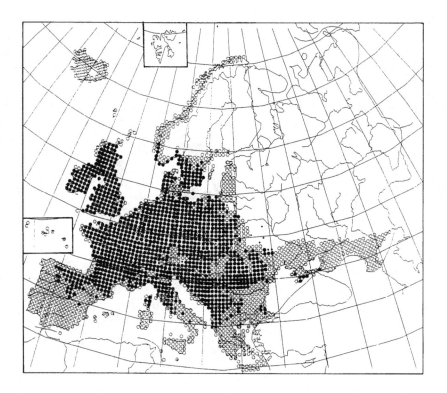

also on the phenological stage and on acclimatisation. Therefore it is necessary to follow the resistance during winter and to examine different overwintering parts to obtain a reliable measure of frost tolerance. In Japan it has been shown that species of warm-temperate elements have a lower frost resistance than elements of colder districts (Sakai & Larcher 1987), and this is also the case in Europe.

One way of studying frost resistance in plants is by differential thermal analysis. The part of a plant to be tested is placed in a container with a thermal sensor inserted in the tissue and gradually cooled. When the temperature drops below zero, ice starts to form in the tissue and this liberates heat which is picked up by the thermal sensor. As the temperature drops further, more ice is formed. This ice is formed between the cells and it leads to a dehydration. But when the temperature falls below –40 °C, ice is formed within the cells in many plants, and this is lethal. George *et al.* (1974) found that temperate trees in North America restricted to areas with winter minimum temperatures that never fall as low as –40 °C consistently gave evidence of ice formation in the cells below –40 °C, while trees growing in areas of colder climate did not. A correlation with

Fig. 17. Distribution of *Cerastium semidecandrum* in relation to the −6 °C isotherm for the coldest month of the year. Filled circles show species presence, open circles the other grid squares of *Altas Florae Europaeae* where the coldest month at the lowest point has −6 °C or more. Points show presence in grid squares outside the indicated isotherm.

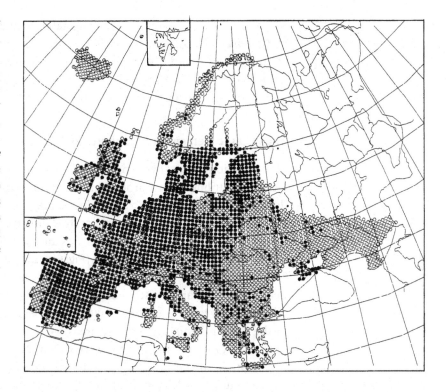

anatomy was also shown. Almost all species with large ring-pores in the xylem tended to be sensitive, while the resistant species, with very few exceptions, had diffuse pores or were non-porous, with narrow lumina in the xylem.

There is also evidence that populations in nature suffer during unusually cold years. Szafer (1932) observed that all beeches on the flood-plains in Poland died during a cold winter in 1923 when the temperature dropped below −40 °C, but survived on the warmer hillsides. The altitudinal limit of beech increases towards the southwest and it forms the climatic timber-line in the western Pyrenees, southern Italy and the Balkans. It is lower in the Alps, Germany and Poland. At high elevations in the western Alps I have observed that branches of beech protruding above the snow were damaged following a cold winter, while lower branches covered by snow were undamaged. *Ilex* populations in northern Jutland in Denmark were exterminated by frost during cold winters in the 1940s (Iversen 1944). *Erica cinerea* in West Norway was damaged during the same winters (personal information from the late R. Nordhagen). During unusually cold winters olive trees in southern France are sometimes severely damaged. In wind-exposed areas in the Scottish Highlands winter browning ('frosting') of *Calluna vulgaris* may occur. It is a desiccation effect resulting from continued transpiration during cold weather when soil moisture is locked in the frozen soil (Watson *et al.* 1966). Browning does not occur if *Calluna* is covered by snow or is growing in woods or other shaded habitats.

The oceanic element of bryophytes and lichens

The oceanic mosses, liverworts and lichens have, in general, a south-western distribution in Europe like the atlantic vascular plants, but the details of this pattern are rather different. While the atlantic vascular plants have distribution patterns closely correlated with winter tempera-tures, the patterns of oceanic bryophytes and lichens more closely follow humidity factors, amount of rainfall (Fig. 6, p. 20), and rainfall frequencies. Thus the areas richest in such bryophytes and lichens are southwest Scandinavia, Ireland, western Scotland and England, the western Alps at altitudes between 600 and 1800 m (Schauer 1965–6), and the western part of the Iberian Peninsula, namely areas without any drought stress even during the driest month of the year (Fig. 7, p. 21). Oceanic plants tend to be absent from the lowlands of the continent and the Mediterranean region, but may be found at higher elevations. Within the areas of high rainfall and humidity in Europe, temperature factors appear also to be important. The southern eu-oceanic species appear to be limited by winter cold, whereas the northern eu-oceanic species may be limited by summer warmth (Ratcliffe 1968). Important contributions to the phytogeography and ecology of this element are given by Degelius (1935), Schauer (1965–6), Greig-Smith (1950), Ratcliffe (1968), and Størmer (1969).

As pointed out by Degelius (1935, p. 226), the atlantic vascular plants are usually endemic to Europe and the Mediterranean. The oceanic, poikilohydric plants usually have a much wider distribution in humid areas in the Macaronesian islands and along the northern Atlantic and northern Pacific shores, and they are often widely distributed in the tropics. Native Amphi-Atlantic vascular plant species with an atlantic distribution are very few, while such distribution patterns tend to be the rule rather than the exception among the oceanic lichens and bryophytes.

This wide difference in distribution patterns between vascular plants and bryophytes and lichens is reasonable because of their different water economies. The stenohydric vascular plants take up water from the soil by means of their roots and hence depend on an adequate water supply in the soil to meet their needs. The relation between humidity factors and the performance of these plants is indirect. It is not easy to establish correlations between the distribution patterns of vascular plants and humidity factors. The poikilohydric bryophytes and lichens take up water directly from the atmosphere and the rain, and if they dry out they go into dormancy and start metabolic activity only when water becomes available again, provided that the drought has not been too severe.

A good example of the difference in the distribution patterns of stenohydric and poikilohydric plants has been provided by Lye (1970). He found the highest concentration of atlantic vascular plants in southwest Norway, close to the open sea where winter temperatures are least severe. The highest concentration of oceanic bryophytes and lichens is found in the zone of maximum rainfall in the fjords. The only vascular plant following this pattern is *Hymenophyllum wilsonii*, a poikilohydric filmy fern. The best correlation with the distribution limits of the oceanic bryophytes that Lye found was the humidity index of Amman (Fig. 18).

In Scotland the highest concentration of oceanic bryophytes and lichens is also on the mainland with maximum humidity and not on the islands with the higher winter temperatures. Ratcliffe (1968) found the number of days per year with precipitation of 1 mm or more to be a good climatic correlative. Hill *et al.* (1991–4) have provided maps of annual rainfall and number of rainy days in the British Isles for comparison with the distribution of bryophytes in Britain and Ireland.

1 The eu-oceanic sub-element

According to Degelius (1935, p. 204) it is possible to subdivide the oceanic element into two sub-elements. The first comprises species that are extremely oceanic and are called **eu-oceanic**. Within this sub-element there are several different groups.

(a) *The southern eu-oceanic group.* Some of the species in this group have a northern limit in the British Isles, for example *Ramalina portuensis* and *Pseudocyphellaria aurata*. Other eu-oceanic plants have a southern distribution but reach the western coast of Norway, for example the lichens such as *Sphaerophorus melanocarpus, Leptogium burgessii, L. hibernicum, Pseudocyphellaria intricata, P. norvegica, Sticta limbata* and *S. canariensis*. Among the liverworts *Acrobolbus*

Fig. 18. Distribution of the poikilohydric liverwort *Pleurozia purpurea* in West Norway in relation to the Amman index of humidity compared with the distribution of the stenohydric vascular plant *Primula vulgaris* in relation to the –2°C January temperature isotherm. After Lye (1970).

Pleurozia purpurea

Primula vulgaris

wilsonii, Adelanthus decipiens, Lejeunea flava and *Microlejeunea diversiloba* (Greig-Smith 1950; Hill *et al.* 1991) are southern eu-oceanic species, whereas *Habrodon perpusillus, Trichostomum brachydontium* var. *littorale, Breutelia chrysocoma* and *Hedwigia integrifolia* are southern eu-oceanic mosses (Størmer 1969). Ratcliffe (1968, Table 2) gives a list of 27 bryophytes belonging to a corresponding southern Atlantic group. All of these have a wide tropical distribution.

(b) *The northern eu-oceanic group.* A northern eu-oceanic group is recognised in Britain comprising 25 bryophytes (Ratcliffe 1968); most of them are also found in western Norway but are rare or missing farther south. Examples include *Anastrophyllum donianum, Mastigophora woodsii* and *Scapania ornithopodioides*. Some re-appear along the western or eastern coast of North America (e.g. *Pleurozia purpurea, Bazzania pearsonii*), and a few have a wide distribution in the Tropics (e.g. *Adelanthus unciformis*). A corresponding element of lichens is especially represented in the forests of Central Norway and re-appears along the coasts of North America. Examples are *Erioderma pedicellatum, Lobaria hallii, Cavernularia hultenii* and *Pannaria ahlneri*. Somewhat similar distribution patterns are displayed by the liverwort *Cephalozia macrostachya* (Hill *et al.* 1991) and the mosses *Andreaea alpina, Oedopodium griffithianum* and, perhaps, *Bryhnia novae-angliae* (Størmer 1969).

(c) *The widespread eu-oceanic group.* Ratcliffe (1968) lists 24 bryophytes that form a widespread eu-oceanic group, including *Dicranum scottianum, Glyphomitrium daviesii, Plagiochila punctata* and *Saccogyna viticulosa*. *Pseudocyphellaria crocata* and *Degelia atlantica* (Fig. 19) might be lichen examples. They rarely appear east of westernmost Norway but may occur in the western Alps and in the Iberian Peninsula.

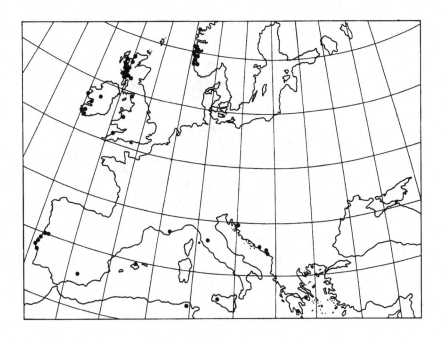

Fig. 19. Distribution of *Degelia atlantica*, a typical widespread eu-oceanic lichen species. After Jørgensen (1978).

2 The sub-oceanic sub-element

The other sub-element, with a wider oceanic distribution, is called **sub-oceanic** and extends eastwards to the Baltic Sea but not as far east as Russia and is widespread in moist areas of Europe. An example is *Sticta sylvatica* (Fig. 20). Other lichen examples include *Bryoria bicolor, Leptogium cyanescens, Lobaria amplissima, L. virens, Nephroma laevigatum, Pannaria conoplea, P. rubiginosa, Platismatia norvegica* and *Degelia plumbea*. Many bryophytes have similar distribution patterns (Størmer 1969), for example *Antitrichia curtipendula, Isothecium myosuroides, Plagiomnium undulatum, Neckera crispa, Pseudoscleropodium purum, Thuidium tamariscinum, Isopterygium elegans, Plagiothecium undulatum* and *Rhytidiadelphus loreus*. Ratcliffe (1968) lists 45 bryophytes in Britain as sub-atlantic species, and in addition several of his list of 24 widespread atlantic bryophytes probably belong here.

Fig. 20. Distribution of *Sticta sylvatica*, a sub-oceanic lichen in Europe. After Degelius (1935).

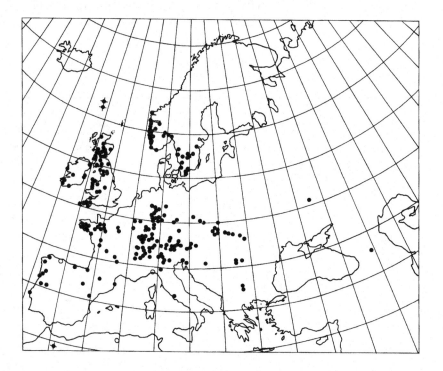

Ecophysiological limitations of poikilohydric plants

Research on the physiological mechanisms limiting the range of poikilohydric species has involved experiments on resistance to drought (Dilks & Proctor 1974, 1976a, b; Proctor 1981). Especially illuminating are the early studies made by Clausen (1952) on liverworts. The vitality of the cells can reliably be tested by recording plasmolysis. Wide differences in drought tolerance were observed, mainly correlated with habitat preferences. Thus the cells of *Cephalozia lunulifolia*, a small species living within tufts of *Sphagnum* in bogs, did not survive 3 hours of desiccation at 97% relative humidity, while *Porella platyphylla*, a species growing on trunks of trees, survived 24 hours desiccation at 0% relative humidity (see also Proctor 1981 for a review of other studies on drought tolerance). Drought resistance may vary at different times of the year (Dilks & Proctor 1976a), with low desiccation tolerance in autumn and winter and increased tolerance in spring and summer.

A good example of the importance of humidity conditions for the survival of oceanic plants is afforded when a forest harbouring oceanic lichens and bryophytes is clear-felled. The humidity conditions are completely changed and the oceanic species die. When the forest grows up again and a canopy is restored the oceanic species come back slowly, if at all. Selective felling damages the flora, but is not nearly as destructive as clear-felling. Gauslaa (1985) found that the richness of oceanic lichens in oak forests in southern Norway was primarily related to forest continuity in the past. Ratcliffe (1968, p. 406) referred to two localities in Wales, one with a rich oceanic bryophyte flora, the other with a poor representation. He suggested that in the rich locality there has been continuous forest or scrub cover since the post-glacial climatic optimum, while in the poor site the forests have been felled and replanted as shown by various alien trees in the forest.

By means of pollen analyses of small hollows within the former woodlands of North Wales, two rich and two poor in oceanic bryophytes, Edwards (1986) showed that the presence and richness of desiccation-sensitive bryophytes depended not only on the extent and nature of woodland disturbance but also on the quality and abundance of available rocky habitats. The two woods richest in rare oceanic bryophytes have had some continuous tree canopy for the last 1500 years *and* have an abundance of damp, shaded habitats such as cliffs, blocks, and boulder-strewn slopes.

Today clear-felling is the preferred silvicultural practice, and this can have devastating effects on humidophilous plants. The lichen *Usnea longissima* used to be quite common in forest tracts north of Oslo and in many parts of Scandinavia (Ahlner 1948) but is now quite rare (Esseen *et al.* 1981; Olsen & Gauslaa 1991), no doubt due to clear-felling. The extinction of the lichen *Erioderma pedicellatum* (Jørgensen 1990) and the serious reduction of *Lobaria hallii* and *Pannaria ahlneri* from the Scandinavian flora are also due to forest felling at or near their sites.

In addition to the overriding importance for oceanic bryophytes and lichens of humidity and drought avoidance, temperature effects may also be important in limiting the northern range of southern eu-oceanic species and the southern extent of northern eu-oceanic species (Ratcliffe 1968). In the case of *Adelanthus decipiens*, a southern eu-oceanic species occurring in western Britain as far north as West Ross, and in Ireland, western France, Iberia, Macaronesia and South America, Ratcliffe (1968) has shown that its distribution within Britain and Ireland is closely delimited by the zone with 190 or more wet days a year (a measure of consistently high atmospheric humidity) *and* by the area with fewer than 23 days a year with snow lying below 65 m (a measure of winter cold). Within the area defined by these two climatic parameters, the local distribution of *A. decipiens* largely depends on the occurrence of old natural or semi-natural woodland with shaded rocky habitats. Moisture and temperature often interact, as temperature affects rates of evaporation. In general the cooler the climate the wetter it becomes. The tendency to cloudiness contributes to the high atmospheric humidity of the atlantic and oceanic climate and helps to reduce diurnal and seasonal temperature fluctuations (Ratcliffe 1968). Interactions of climatic variables are of considerable ecological importance. When climatic factors have been considered individually, it is important to consider them in combination and to relate evidence of their interaction to local field conditions (Forman 1964).

6   The thermophilic element

Introduction

The fact that low summer temperatures restrict the distribution of plants is an everyday experience. Gardeners who try to grow exotic species find that they will not set ripe fruits except in unusually warm summers. After unusually cold summers buds and fruits may fail to ripen and are damaged by early frost. Such early frosts have sometimes destroyed the grain harvest over large areas, resulting in famine. By comparing the latitudinal distribution of cultivated species in Europe a characteristic pattern of distribution limits is found towards the north. Some species are restricted to the southernmost parts of the area with the warmest summers. Typical examples are the traditional cultivation of grapes, going north to the Rhine Valley in Germany, into southern England but not reaching northern England; and the cultivation of maize north to central England and southern Sweden, of wheat north to Scotland and in the valleys of southern Scandinavia north to Trondheim in Central Norway, and potatoes which can be grown all over the British Isles and north to the inner fjords of North Norway.

Summer temperatures limit the distribution of species altitudinally as well as latitudinally. The altitudinal limits of plants in the Alps tend to be higher than in Fennoscandia. This is shown in Fig. 21, where the altitudinal limits of plants in southern Norway are plotted against the same species in the Alps. It is seen that the altitudinal limits in the Alps are, on average, 1000 m higher than in southern Norway. The climatic timber-line in Tyrol is at 2300 m and in southern Norway at 1250 m. Permanent farms are found up to 1800 m in the Alps and to 750 m in southern Norway.

Many indices have been constructed to correlate with the performance of thermophilic plants; these have been reviewed by Tuhkanen (1980). The best known is probably the heat sum. This assumes that a minimum sum of temperatures above a certain threshold temperature (measured in degree-days) is needed for cultivated plants to complete their growth cycle. Often the threshold temperature is taken to be +5 °C. This index is much used in agricultural meteorology. Many tree species follow a similar pattern.

Fig. 21. Altitudinal limits of plant species in southern Norway compared to their corresponding limits in the Alps. The data from southern Norway are from Lid (1985) and Hultén (1971a) supplemented by information provided by F. Wischmann, Botanical Museum, University of Oslo. The altitudinal limits for the plants in the Alps are the means from Tyrol (based on Dalla Torre & Sarntheim (1906–12), Klebelsberg (1913), Reisigl & Pitschmann (1958)), from Graubünden (based on Braun-Blanquet & Rübel 1932–5), and from Wallis (based on Jaccard (1895) and Becherer (1956)). Only species that are fairly common in both areas are included.

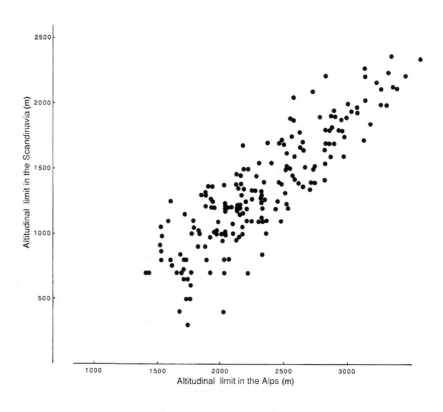

Timber-lines

One of the most conspicuous vegetational limits in the world is the alpine and arctic timber-line. The division between alpine and subalpine and arctic and subarctic is often drawn at the timber-line. Important information about timber-line observations is found in Hermes (1955), Plesnik (1971), Troll (1972), Zwinger & Willard (1972), Ellenberg (1978), Franz (1979), Tranquillini (1979) and Dahl (1986).

Different kinds of timber-lines can be considered (Hustich 1966). The seemingly simplest is the empirical or actual timber-line, that is the uppermost or the latitudinal limit of forest that can be depicted on a topographical map. To be more precise, definitions must be given to indicate what constitutes a tree, and the number of and the distances between the trees to constitute a forest (Dahl 1957, 1986).

The actual timber-line is, from the ecologist's point of view, highly complex, because so many different local ecological factors interact in its formation. One locality may be too exposed to wind, another may have impeded drainage, a third can have too prolonged snow-cover in the spring for any forest development. Very often the actual timber-line is also highly influenced by man's use of the land by felling trees for fuel, whereas grazing by sheep, cattle and goats often prevents regeneration of the forest. According to Ratcliffe (1977) a true, natural timber-line is rarely observed in Britain due to the long-term effects of man and his animals.

Another type of timber-line, generally called the climatic timber-line (Dahl 1986), is more amenable to ecological analysis. As one ascends a mountain, the treeless areas expand at the expense of forests and woods – until one reaches a limit where the last woodlands or forests disappear. As altitude increases, temperature decreases and reaches a limit where the climate no longer can support the growth of forests, even under favourable local conditions. In the Northern Hemisphere, and in areas with reasonable moisture in summer, the uppermost forests are normally found in protected valleys with southern exposure and with little snow in winter. A close approximation to the climatic timber-line, defined as the altitudinal limit of forest growth under optimal local conditions, is found by recording the highest forest within an extended area, for example the area of a map-sheet at the scale of 1:50 000. By examining such maps, and using proper ground-control, the regional variation of the climatic timber-line can be mapped – a method invented by Imhof (1900). Some qualifications are necessary. A variety of ecological opportunities at high altitude must be available; hence no climatic timber-line can be recognised unless there are, in the neighbourhood, mountains that exceed the climatic timber-line by 200 m. Figure 22 gives isohypses for the climatic timber-line in Scandinavia. It is seen that it reaches its maximum in the highest mountain massif of Jotunheimen at 1230 m. From there it drops, most sharply towards the west and more gently towards the east and north. The poleward coastal fringe of northern Norway is north of the arctic timber-line. In Scotland the highest timber-line is at 640 m in the western Cairngorms. It is, however, surprising that it is pine and not birch that forms the timber-line there. The timber-line drops towards the north and west to probably 300 m inland in northwest Sutherland (Ratcliffe 1977, p. 73). In the Hebrides, the northern part of Lewis is treeless today and was probably never densely forested during the post-glacial, although local copses were present according to Birks (1988, 1991).

Climatic correlations of the timber-line have often been discussed. In general it is agreed that summer temperature is a limiting factor, although the actual physiological mechanisms involved are still under discussion.

Fig. 22. Isohypses of
climatic timber-lines in
Scandinavia. Altitudes
are in metres. After
Abrahamsen *et al.*
(1977).

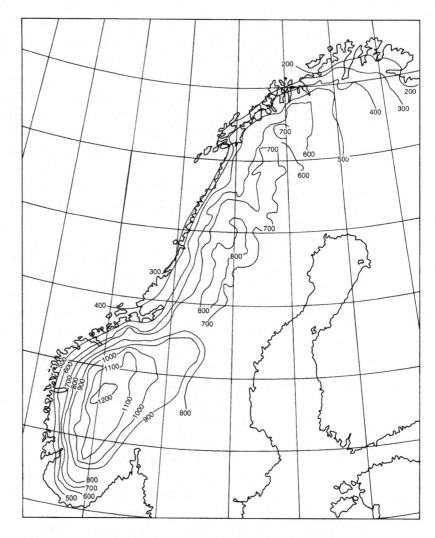

Köppen (1920) thought that the timber-line correlated with a July
temperature of +10 °C. Mayr (1909) introduced the mean of the four
warmest months, the tetratherm, as a better climatic measure; this is close
to the mean of the vegetation season. Aas (1964; see also Dahl 1986) found
that the mean of the three warmest months, the tritherm, correlated better
than the tetratherm for timber-lines in Norway, an observation confirmed
by data from the western USA (unpublished observations). Quervain
(1903) and Brockmann-Jerosch (1919) observed that timber-lines tended
to be high in areas of high mountain massifs and introduced the
'*Massenerhebung*' as a factor.

Consistent with the original findings of Hagem (1917), Aas (1964) found that the tritherm temperatures at timber-lines were consistently higher in oceanic than in continental parts of Norway. This seems also to apply on a global scale in regions with a seasonal climate. Figure 23 plots the tritherms at the timber-line level against a continentality index. As a measure of continentality the annual amplitude (difference of the mean temperatures of the warmest and coldest month) corrected for latitude by dividing it by the Conrad factor of latitudes, $\sin(\phi + 10°)$ (where ϕ is the latitude) has been used.

It is seen that the points in Fig. 23 follow a consistent pattern with a slight negative correlation between the tritherm and the continentality index. A consistency, in the first place, is highly surprising since the species forming the timber-line are different. In Scotland it is *Pinus sylvestris*, in Norway *Betula pubescens* ssp. *tortuosa*, in the French Alps *Pinus uncinata*, in the Tyrol *Pinus cembra* and *Larix decidua* (sometimes also *Picea abies*), in Colorado *Picea engelmannii*, in Nepal *Betula utilis*, in Australia *Eucalyptus nipophila*, and in New Zealand *Nothofagus solanderi* and *N. menziesii*. A correlation thus exists between a life form (tree) or a formation or biome (forest) and summer temperature,

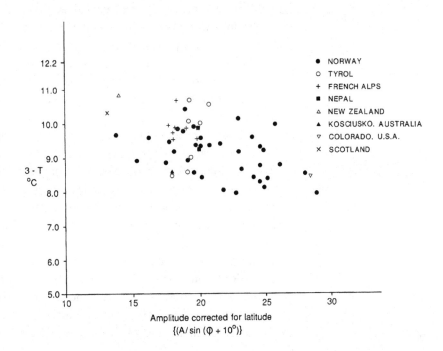

Fig. 23. Tritherm temperatures (3-T°C) at timber-line plotted against continentality. After Dahl (1986) with the addition of a timber-line observation in Scotland, according to Ratcliffe (1977) and meteorological data from the station at Dalwhinnie.

regardless of the genetic stock from which the timber-line tree species evolved. This was already observed by Daubenmire (1954) who remarked:

> Because a great many genetically distinct trees contribute different segments of a timber-line pattern that has a remarkable geographic conformity, the hypothesis is suggested that a major ecological principle is involved that may be analogous to the wilting coefficient, in which some environmental complex abruptly exceeds the tolerance of all trees regardless of the variation between them.

The important question to answer is which ecological principle is involved, i.e. which physiological processes prevent trees from adapting themselves to summer temperatures below a certain limit. Such a physiological process must be of a very general applicability, since it must apply to deciduous as well as evergreen species and to gymnosperms as well as angiosperms. The process should also account for the slightly negative correlation between limiting temperatures and continentality.

Respiration and growth

The many indices used for correlation between the northern and alpine limits of plants and meteorological factors are rarely based on any ecophysiological foundation. The question is, in which way does temperature affect plant performance? At low temperatures growth is low or zero, but it increases rapidly with increasing temperatures. Growth is not a linear function at low temperature but is a curve with an upward curvature until the growth reaches a maximum. Is there an explanation for this?

It is possible to think of an actively growing plant as a factory (Fig. 24). It has a raw materials department that utilises solar energy, carbon dioxide, water and minerals to produce organic components: carbohydrates, amino acids and fats. These are raw materials which are transformed during the growth process, especially in the meristems, to new tissue. But the chemical process that converts raw materials to finished products requires additional energy in the form of adenosine triphosphate (ATP). The ATP is produced by dark respiration in the mitochondria. We can consider dark respiration as an energy department which provides the energy needed for the growth processes. Energy and raw materials are both necessary for growth. Besides energy and raw materials for growth, some energy is needed for maintenance and transport. Dark respiration is strongly

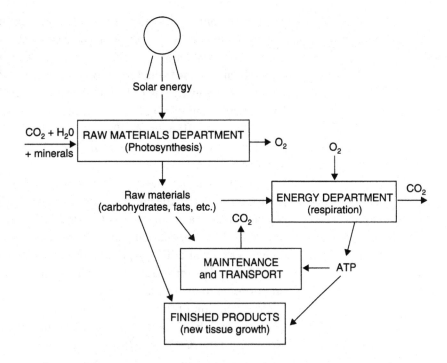

Fig. 24. The plant production system. Further explanation, see text. After Dahl (1986).

influenced by temperature. A working hypothesis is that the energy supply from respiration is the rate-limiting factor for growth at low temperatures.

The relation between respiration and growth has been studied by de Vries and co-workers (de Vries 1974, 1975; de Vries *et al.* 1989). Let us first take a simple physiological process in plants, the synthesis of the amino acid lysine from glucose and ammonia. The equation is

$$1 \text{ glucose} + 2 \text{ NH}_3 + 2 \text{ NADH}_2 + 2 \text{ ATP} =$$
$$1 \text{ lysine} + 4 \text{ H}_2\text{O} + 2 \text{ NAD} + 2 \text{ ADP} + 2\text{P}.$$

One glucose is rebuilt to lysine, but in the process 2 NADH_2 and 2 ATP are consumed. These substances are produced by dark respiration at the expense of glucose. Thus more than one molecule of glucose is consumed for the production of one molecule of lysine. Summing up, the end result is

$$1 \text{ g glucose} + 0.156 \text{ g NH}_3 + 0.039 \text{ g O}_2 =$$
$$0.691 \text{ g lysine} + 0.255 \text{ g CO}_2 + 0.269 \text{ g H}_2\text{O}.$$

The combustion value of transferring 1 g glucose to 0.691 g lysine involves an energy loss by respiration of 35%. In this case some of the ATP can be produced by photosynthetic phosphorylation in the leaves but, by far, most of the ATP needed for growth must be produced in the meristems.

Similar calculations can be made for other products. The most economical process is the formation of starch and cellulose from glucose where the energy loss is only 7%. Proteins require much energy, about 66%, fat 20%, and lignin 26%. To produce an average maize or corn plant involves a respiration loss of 20% (de Vries 1974, 1975). In general, crops producing carbohydrates have higher yields than those producing proteins.

In a meristem of maize which is supplied with carbohydrates but produces the ATP in the meristem, one new gram of maize tissue cannot be produced before an equivalent of 0.2 g glucose is respired. The time it takes to respire the corresponding amount of glucose is necessarily dependent on temperature and the time it takes is inversely related to the respiration rate. Thus, at low temperatures, the efficiency of temperature for growth is proportional to the respiration rate.

Confirmation of this idea has been obtained by studying the growth and respiration of apical shoots of spruce in a subalpine forest in Norway (Dahl & Mork 1959). Here moisture is abundant and daily solar radiation high, but the temperature is low. The daily growth of the shoots increases with increasing temperature. A curve of the respiration in the growing shoots as a function of temperature is shown in Fig. 25. Temperature was measured by means of a thermograph, and respiration was quantified by Warburg techniques. The respiration sums or respiration equivalents (Re-sums) are the values appearing when the temperatures, as shown by the thermograph, are weighted according to their effect on respiration, using 1 day respiration at 10 °C as a unit. The respiration sums are compared with plant growth in Fig. 26. It is seen that the apical growth of spruce is highly correlated with the respiration sums, with a correlation of 0.98.

The fitted line cuts the zero growth axis at Re = 9.4 which is the basal respiration required for maintenance and transport. It corresponds to a respiration during 24 hours at 2.8 °C and this is called the basal temperature (Skre 1972). Only respiration in excess of the basal respiration will result in growth. It is concluded that plant growth is limited by dark respiration or some factor closely correlated with dark respiration. Re-sums (= respiration equivalents) are a synthetic measure of the potential for plant growth and ripening.

This model may explain why plants need higher mean temperatures

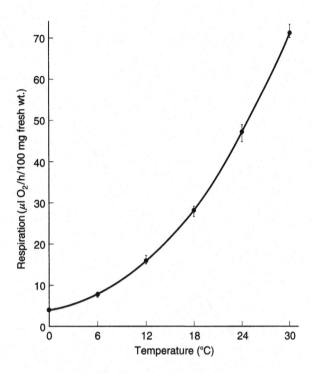

Fig. 25. Dark respiration of apical shoots of Norway spruce (*Picea abies*) as a function of temperature. After Dahl & Mork (1959).

during the growing period in oceanic than in continental areas. In continental areas the temperature amplitudes are larger than in coastal areas. From Fig. 25 it is seen that at a given mean temperature, high amplitudes result in higher Re-sums than low amplitudes because of the upward curvature of the respiration curve. Hence the plants can survive at lower mean temperatures in the growing season in a continental area than in coastal areas because the incidences of high temperature in continental areas result in a relatively higher output in terms of respiration.

The limitation of temperature on growth is of a different kind than the limitation of light and water on photosynthesis. Photosynthesis is a source of materials, while growth is a sink. If the source capacity exceeds the sink capacity some of the production becomes superfluous. The plants require means to dispose of excess energy. Perhaps photo-respiration is one such process; cyanide-resistant respiration may be another (Lambers & Rychter 1989; Robinson *et al.* 1992). The plants can also release absorbed energy by fluorescence.

There is evidence that crop production at least in temperate areas is positively correlated with temperature, suggesting that the sink capacity is limiting. According to, for example, McCree (1970) and Kimura *et al.*

Fig. 26. The relation between the daily apical growth of spruce (*Picea abies*) in a subalpine forest of Norway and the sum of temperatures weighted according to their effect on respiration (Re-sums). After Dahl & Mork (1959).

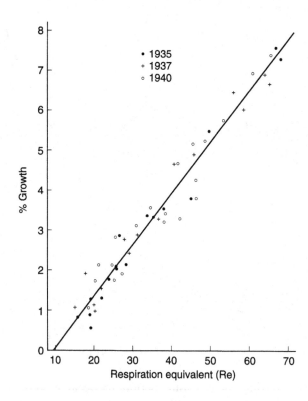

(1978), a positive linear relationship exists between dark respiration rates and relative growth rates in plants, above a certain maintenance respiration.

What possibilities do plants have to adapt to low temperature limitations? One way is to increase the respiration, for example by a higher number of mitochondria in the cells. Miroslavov & Kravkina (1991) found by counting mitochondria that the leaf chlorenchyma cells of *Poa alpina* from 3450 m altitude had about twice as many mitochondria in the chlorenchyma cells as corresponding cells of the same species from 2600 m. Similar results were also obtained for *Oxytropis lapponica*. In general, plants from cold climates have a higher respiration than corresponding plants from warmer climates when measured at the same temperature (Larcher 1984, p. 134). Billings & Mooney (1968) found that plants from high-altitude stations have a higher rate of respiration than comparable plants from lower levels when measured at the same tempera-ture. But in nature, the total respiration was approximately the same because temperatures at higher levels were lower. This makes good ecological sense. Respiration above what is needed for growth is a loss since it is at the expense of net primary production. On the other hand, with low respiration rates, temperatures must be higher to exceed the basal respiration and to initiate growth in spring and early summer.

Calculation of respiration sums

Respiration sums for the lowest altitudes in each square in the *Atlas Florae Europaeae* system have been calculated to permit comparisons with the distributional patterns of thermophilic plants. The calculations are explained in Appendix I. They are based on the physiological properties of spruce with a basal temperature of 2.8 °C and an activation energy of respiration of 70 kJ/mol. This corresponds to a temperature coefficient Q_{10} of 2.65 (Q_{10} is the ratio between the dark respiration rates at 20 °C and 10 °C). The respiration in one month at 10 °C is taken as one unit. The model is, strictly speaking, only applicable to spruce. However, the basic physiological parameters, for example the Q_{10}, are not very different for different species, varying between 2.0 and 3.0. Spruce can be taken to be fairly representative of continental species, but less representative of oceanic species. Effects of other values of the parameters are discussed on p. 75.

Instead of the respiration sums (Re) its logarithmic transform $R = \ln Re$ is used. The reason for this is that the R-values vary approximately linearly with altitude; a 100 m altitudinal difference corresponds to a difference in R in southern Norway of 0.132 per 100 m between 0 and 1000 m, in continental Scotland the gradient is 0.085 per 100 m. This permits a direct comparison between latitudinal and alti-tudinal limits.

The Iversen thermal-limit diagrams

Iversen (1944) devised a diagram to define and delimit the **thermosphere** of individual species. As one measure he used the mean of the warmest month on an inverted scale along the vertical axis and the temperature of the coldest month on an inverted scale along the horizontal axis. The diagram for *Viscum album* (Fig. 27) is shown as an example. Stations where *Viscum* is present are represented by filled circles.

From this diagram we can see that the set of localities is limited by a vertical line at about –8 °C. This suggests that *Viscum* is sensitive to winter frost. But at winter temperatures higher than –8 °C, the thermosphere is

Fig. 27. Iversen diagram of the thermosphere of *Viscum album*. Along the horizontal axis the mean temperature for the coldest month (T_1) is plotted, along the vertical axis the mean temperature for the warmest month (T_7) is plotted. The distribution limit corresponds with the isoline representing an accumulated annual respiration of 6.8 Re-units (R = 1.95). After Skre (1979a).

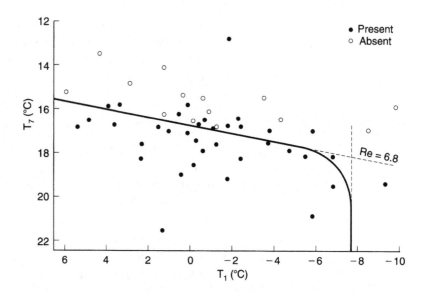

limited by an oblique line. Obviously summer temperature also limits the growth of *Viscum* in combination with winter temperature. The falling slope reflects the fact that a plant needs a higher summer temperature to survive in a continental locality with a short growing season and low winter temperatures than in an oceanic locality with a long growing season and high winter temperatures.

The technique has also been used by Hinttikka (1963) with a number of combinations of climatic parameters. Besides winter and summer temperatures, humidity factors were also considered. Figure 28 presents his diagram for *Tilia cordata*. Apparently *Tilia* is not limited by winter temperature but has an oblique limitation towards colder summer temperatures.

The question arises whether the respiration hypothesis can explain the oblique limitation of the thermosphere of *Viscum album* and *Tilia cordata*. Clearly, at a given July temperature the heat sum will increase with increasing winter temperatures because the duration of the growing season increases.

Skre (1979a) found that the winter and summer temperatures in combination were good predictors of the accumulated annual respiration (the respiration sum) and that in an Iversen diagram the lines of constant respiration sums corresponded with the distribution limit and had a decline inversely proportional to the temperature coefficient Q_{10} of dark

Fig. 28. Iversen diagram of the thermosphere of *Tilia cordata* according to Hinttikka (1963). The distribution limit corresponds with the isoline representing an accumulated annual respiration of 5.4 Re-units (R = 1.7). After Skre (1979a).

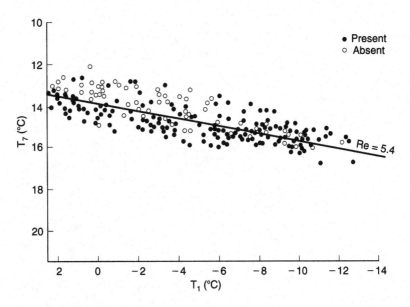

respiration in spruce. For *Viscum* this limit corresponded to a constant R of 1.95 (Re = 6.8). In the distribution map of *Viscum* (Fig. 32, p. 71) it is seen that the present occurrence of *Viscum* is limited by the 2 R-unit isoline of respiration sums. *Viscum* has, however, been found as Holocene fossils outside its present distributional limits, and the limits of subfossil *Viscum* pollen correspond to a 1.8 R-unit isoline. This suggests that the climate was once 0.2 R-units warmer than today. The thermosphere of *Tilia cordata* (Fig. 28) is limited by the 1.7 R-unit (Re = 5.4) and its present-day distribution corresponds to the 1.6 unit isoline (Fig. 35, p. 74).

Correlations between distribution limits and respiration sums (R-values)

Figure 29 gives an isoline map of the R-values calculated for the lowest point in each of the *Atlas Florae Europaeae* mapping squares. For all species that have adequate available distribution maps, the distribution limits were compared with these R-isolines. The results are given in Appendix II. Based on such comparisons a set of thermophilic sub-elements can be recognised, as discussed below.

The preference of species for various summer temperatures is expressed by Ellenberg's '*Temperaturzahl*' (Ellenberg *et al.* 1991), based on the distribution patterns of wild plants in Central Europe. This is a 10-point scale where 1 represents species growing at high-alpine and mid-alpine

Fig. 29. Isolines of R-values for Europe based on spruce respiration calculated for the lowest point of each of the *Atlas Florae Europaeae* squares.

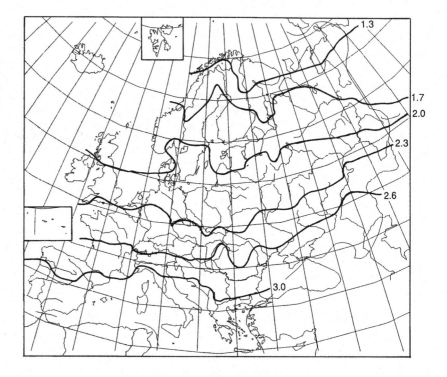

climates, 2 represents species preferring low-alpine and subalpine climates, and so on down to 9 for species restricted to Mediterranean climates. Obviously there must be a relation between the limiting R-values and Ellenberg's *Temperaturzahl*. For the 426 taxa where R-values and T-values are both available, the correlation between them is 0.62, which is highly significant, but with a considerable variation. This is mostly because the R-values measure the tolerance of the species to cold climates expressed as altitudinal or latitudinal limits, whereas the T-values express the preference of the species for certain temperature regimes where the lower limits are also important. Species with a broad altitudinal distribution are very often classed 'indeterminate' in the Ellenberg system.

In a map depicting the vegetation of Europe, produced for conservation purposes (Noirfalaise 1987), the limits between the *vegetation regions* are seen to correspond broadly to the isolines in Fig. 29. In the vegetation map the following major regions are recognised.

1 *A Mediterranean biome* which is further subdivided into a thermo-mediterranean zone, a meso-mediterranean zone, and a supra-mediterranean zone.

2 *A temperate biome* dominated by deciduous trees, with *Quercus* and *Fagus* as dominants. This stretches north to Denmark and southernmost Sweden and Norway, and in Britain north to southern Scotland where *Quercus* is a climax tree (McVean & Ratcliffe 1962).

3 *A boreal biome* with conifers as the dominant trees, delimited northwards and altitudinally by the climatic timber-line. This biome is further subdivided by Scandinavian authors (Påhlsson 1984; Dahl *et al.* 1986). In a *boreo-nemoral region* in the south, oak is present in climatically favourable localities but spruce and pine are the climax trees. It covers southernmost Finland, Sweden as far north as just north of the large Swedish lakes, and parts of southeast Norway with outliers in the western fjords north to Trondheim. It also covers the lowlands of northern Scotland with limited oak-woods and with pine (McVean & Ratcliffe 1962). There is then a *southern boreal region* with spruce and pine, but with scattered occurrences of temperate deciduous trees and of agriculture based on barley and wheat. Above and north of this region is the *middle boreal region* with agriculture based on potatoes and milk production. Beyond this, there is the *northern boreal region* up to the climatic timber-line and with only summer farming.

4 *An arctic-alpine biome* above and north of the timber-line. In its lower and southern part the vegetation is dominated by shrubs, *Betula nana*, *Juniperus communis* and *Salix* spp. This is the *low-alpine region* up to about 1400–1500 m in southern Norway, 1100 m in northern Sweden, and about 1000 m in Scotland. Above this is the *mid-alpine region* with grasses and herbs such as *Juncus trifidus*, *Festuca ovina*, and *Carex bigelowii*. Higher still is the *high-alpine region* where all the soils are solifluction soils. This region is hardly present in Scotland.

From comparisons of the R-values the following *floristic sub-elements* can be distinguished.

1 *A Mediterranean sub-element* with limiting R-values of 2.4 or higher. Typical examples are *Selaginella denticulata* (Fig. 30), *Pinus halepensis*, *P. pinea*, *Juniperus oxycedra* and *J. phoenicia*, native evergreen oaks, *Spartium junceum*, *Cotinus coggygria*, *Laurus nobilis*, *Pistacia* spp., *Lavandula stoechas*, etc.

2 *A southern temperate sub-element* not going further north than to southernmost Denmark and southern England and limited by R-values of 2.1 or higher. Examples include *Quercus pubescens* (Fig. 31), *Buxus sempervirens*, *Equisetum ramosissimum*, *Populus nigra*, *Nigella*

arvensis, Robinia pseudacacia, Medicago minima, Euonymus europaea, Tilia platyphyllos, Viburnum lantana, Luzula forsteri and *Cephalanthera damasonium.*

3 *A temperate sub-element* going as far north as just north of the Middle Swedish lakes (*limes norrlandicus*), to southernmost Finland and eastwards into Russia, in Norway limited to southeastern parts and with scattered occurrences in the fjords along the west coast north to Trondheim, and in the British Isles occurring north to southern Scotland. The species are limited by R-values of 1.8 or higher. Examples are *Viscum album* (Fig. 32) and *Quercus robur* (Fig. 33). Other examples include *Taxus baccata, Polygonum minus, Aquilegia vulgaris, Cardamine bulbifera, Filipendula vulgaris, Crataegus* spp., *Malus sylvestris, Geranium sanguineum, Cornus sanguinea, Fraxinus excelsior, Ligustrum vulgare, Lathraea squamaria, Jasione montana, Festuca altissima* and *F. gigantea, Bromus benekenii* and *B. ramosus, Melica uniflora, Brachypodium sylvaticum, Cladium mariscus, Carex sylvatica, C. pseudocyperus* and *Epipactis palustris.* Many of the above species are typical of rich mesic deciduous forests.

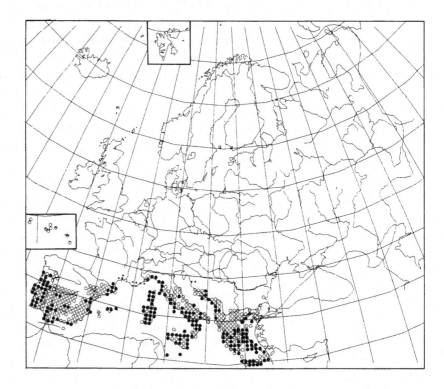

Fig. 30. Distribution of *Selaginella denticulata* in relation to the 3.0 R-value isoline. Filled circles show species presence, open circles the other grid squares of *Atlas Florae Europaeae* where the R-value is 3.0 or higher. Points show presence in grid squares outside the indicated isoline.

Fig. 31. Distribution of *Quercus pubescens* in relation to the 2.3 R-value isoline. In Russia it is also limited by winter frost, by the –3 °C isotherm for the coldest winter month. Filled circles show species presence, open circles the other grid squares of *Atlas Florae Europaeae* where the R-value is 2.3 or higher. Points show presence in grid squares outside the indicated isoline.

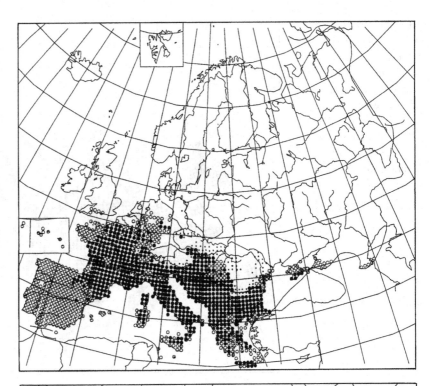

Fig. 32. Distribution of *Viscum album*, compared with the 2.0 R-value isoline and with the –8 °C isotherm for winter temperature in Russia. Filled circles show species presence, open circles the other grid squares of *Atlas Florae Europaeae* where the R-value is 2.0 or higher. Points show presence in grid squares outside the indicated isolines.

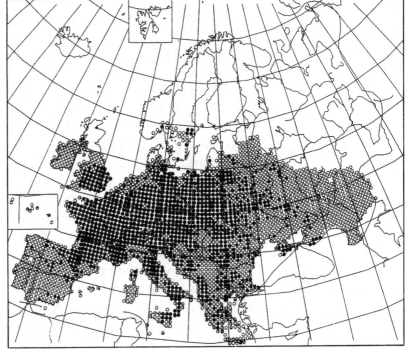

Fig. 33. Distribution of *Quercus robur* in relation to the 1.8 R-value isoline. Filled circles show species presence, open circles the other grid squares of *Atlas Florae Europaeae* where the R-value is 1.8 or higher. Points show presence in grid squares outside the indicated isolines.

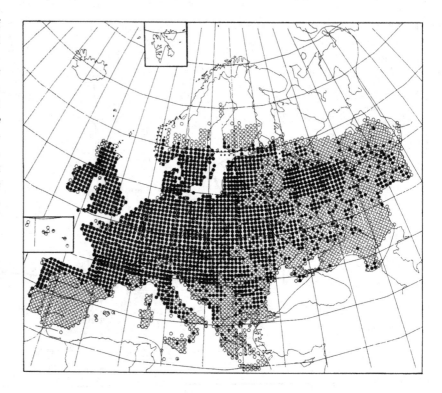

Comparing the limiting R-values in Scandinavia and in Britain, the limiting R-value is commonly found to be higher in Britain than in Scandinavia. There are 72 species where the R-values are estimated separately for the British Isles and Scandinavia and the difference is, on average, 0.13 units or corresponding to a difference in altitude of a little more than 100 m. This is possibly an effect of continentality with a stronger differentiation between local climates in continental parts of Scandinavia than in the more coastal climate in Britain, just as alpine timber-lines are affected northwards and westwards.

4 *A southern boreal sub-element* extending to about 300 m altitude in southern Norway and going north along the Norwegian coast as far as to the inner fjords of the county of Nordland, occurring along the northernmost shores of the Gulf of Bothnia, and covering south Finland. It is limited by R-values of 1.5 or above. In continental Scotland the corresponding altitudinal limit is at about 350 m.

Examples of this sub-element are *Ulmus glabra* (Fig. 34) and *Tilia cordata* (Fig. 35). Other examples include *Polystichum braunii, Corylus avellana, Alnus glutinosa* (Fig. 36), *Humulus lupulus, Silene nutans,*

Fig. 34. Distribution of *Ulmus glabra* in relation to the 1.7 R-value isoline. Filled circles show species presence, open circles the other grid squares of *Atlas Florae Europaeae* where the R-value is 1.7 or higher. Points show presence in grid squares outside the indicated isoline.

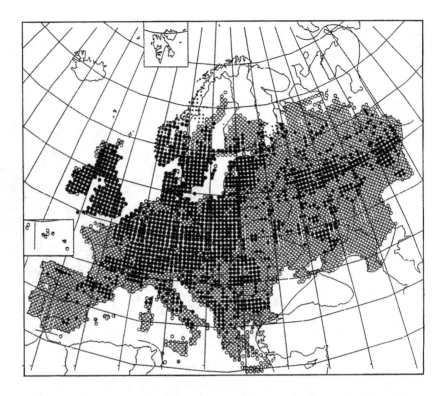

Hepatica nobilis, Ranunculus polyanthemos, Sedum telephium, Geum urbanum, Trifolium medium, Acer platanoides, Impatiens noli-tangere, Lysimachia vulgaris, Acinos arvensis, Origanum vulgare, Solanum dulcamara, Verbascum thapsus, V. nigrum, Viburnum opulus, Lonicera xylosteum, Cirsium vulgare, Convallaria majalis, Polygonatum odoratum, Iris pseudacorus, Briza media, Scirpus sylvaticus and *Rhynchospora fusca.*

5 *A middle boreal sub-element* with limiting R-values of 1.3 or higher. An example of this is *Pteridium aquilinum* (Fig. 37). Other species with similar R-limits include *Arenaria serpyllifolia, Stellaria uliginosa, Anemone nemorosa, Lychnis viscaria, Geranium robertianum, Frangula alnus, Viola riviniana, Scutellaria galericulata, Linaria vulgaris, Cirsium palustre, Eriophorum latifolium* and *Platanthera bifolia.* Several weeds that follow agriculture also have similar limits. The corresponding altitudinal limit in Scotland is at 500 m.

6 *A northern boreal sub-element* which cannot readily be depicted on the broad-scale map of R-values at the lowest altitude in the squares based on the *Atlas Florae Europaeae.* Values must either be correlated with a

Fig. 35. Distribution of *Tilia cordata* (after Pigott 1991) in relation to the 1.6 R-value isoline. Filled circles show species presence, open circles the other grid squares of *Atlas Florae Europaeae* where the R-value is 1.6 or higher. Points show presence in grid squares outside the indicated isoline.

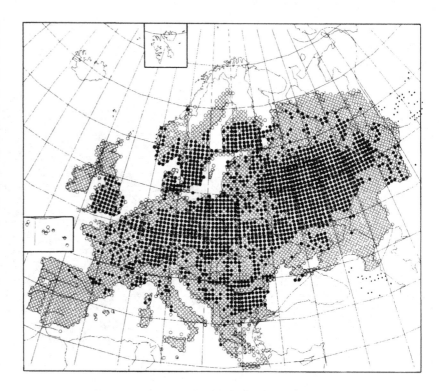

more detailed mapping in Scandinavia (Skre 1979a) or be based on the altitudinal limits of the species. Examples are *Betula pubescens* ssp. *tortuosa* (R = 1.0), *Pinus sylvestris*, *Picea abies* (R = 1.1), *Salix caprea* and *S. pentandra*, *Populus tremula*, *Lychnis flos-cuculi* (Fig. 38), and *Actaea spicata*.

For alpine species the R-values cannot be calculated from horizontal dot-maps, but instead R-values for different altitudes can be computed and compared with the altitudinal limits of the species in different regions. Several mountain tracts have been studied intensely to record the altitudinal limits of species. Skre (1983) used data from 6 such areas in Scandinavia to compute averages of the R-values. Dahl (1992b) lists R-values based on the altitudinal limits of alpine plants in Fennoscandia.

Fig. 36. Distribution of *Alnus glutinosa* in relation to the 1.5 R-value isoline. Filled circles show species presence, open circles the other grid squares of *Atlas Florae Europaeae* where the R-value is 1.5 or higher. Points show presence in grid squares outside the indicated isoline.

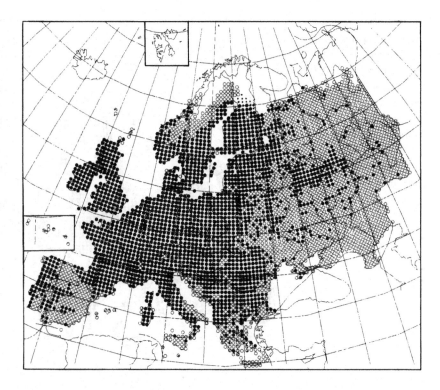

Effect of changing parameters

Constants derived from the study of apical growth of spruce in a subalpine area in Norway are used in the model. These are the activation energy of 70 kJ/mol corresponding to a temperature coefficient Q_{10} of 2.65 and a basal respiration corresponding to respiration at a constant temperature of 2.8 °C.

Vascular plants in general have Q_{10} values of around 2.5. But this value differs between different species and even between different ecotypes of one species. Björkman & Holmgren (1961) found that oceanic ecotypes of *Solidago virgaurea* had a lower Q_{10} than continental ecotypes. The value of Q_{10} might affect the adaptation of species to climate and the different Q_{10} values might result in different distribution patterns.

Figure 39 (p. 78) shows the respiration curves of two species, *Picea abies* with a Q_{10} of 2.65 and *Ulmus glabra* with a Q_{10} of 2.0 but with curves normalised to a unit value at 10 °C. It is seen that below 10 °C *Ulmus* has a higher respiration than *Picea*, while *Picea* respires more above 10 °C. Thus *Ulmus* appears to utilise low temperatures relatively better than

Fig. 37. Distribution of *Pteridium aquilinum* in relation to the 1.4 R-value isoline. Filled circles show species presence, open circles the other grid squares of *Atlas Florae Europaeae* where the R-value is 1.4 or higher. Points show presence in grid squares outside the indicated isoline.

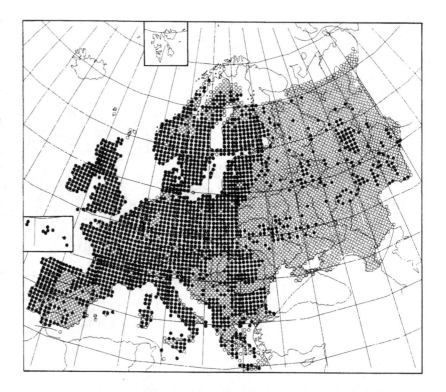

Picea and would therefore have an advantage in oceanic climates with a long and cool growing season, whereas *Picea* would have an advantage in continental climates with a rapid transition between winter and summer and high midsummer temperatures. In an Iversen thermal-limit diagram the tilt of the limiting curve would become steeper in *Ulmus* than in *Picea* due to its lower Q_{10} values (Skre 1979b). In Fig. 40 (p. 79) Iversen diagrams of *Picea* and *Ulmus* are compared. The distribution limits of the two species correspond well with the oblique lines representing constant annual respiration sums of 2.85 units (= R-value of 1.05) for *Picea* and 3.42 (= R-value of 1.23) for *Ulmus*, which are sums based on actual measurements for growing spruce shoots and leaf discs of elm (Skre 1979b). The good correspondence between the actual and the theoretical distribution limits supports the main hypothesis that dark respiration is an important process limiting growth at low temperatures. It also means that species with a low Q_{10} do better in coastal areas and those with a high Q_{10} do better in continental climates (Skre 1993). Comparison of the distribution of *Tilia cordata* (Fig. 35) and *Ulmus glabra* (Fig. 34) brings this out. A similar tendency is also visible in the map of *Quercus robur* (Fig. 33).

Fig. 38. Distribution of *Lychnis flos-cuculi* in relation to the Re-limit of 3 corresponding to R = 1.1. Main area hatched and isolated stands shown as circles. After Skre (1979a).

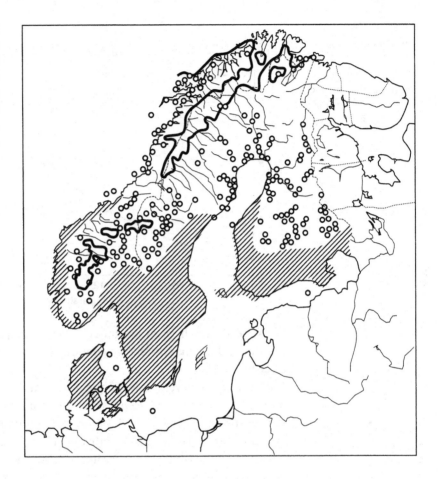

Effects of low summer temperatures on plants

One common effect of low summer temperatures is that fruits and seeds do not ripen. This applies to agricultural crops referred to above and also to wild plants. Pigott (1981) found that the seeds of *Tilia cordata* did not ripen during a series of cold summers near its northern limit in Finland. In northern England, however, *Tilia* did not produce seeds because fertilisation failed (Pigott & Huntley 1981). This might be the reason why it does not extend further north in Britain (Fig. 35). For *Pinus sylvestris* in Norway, Hagem (1917) found that a tetratherm above 10.5 °C was needed to obtain ripe seeds, and that in northern Norway during the last century such summer temperatures limited reproduction to a few years. Eide (1932) and Mork (1933) found that *Picea abies* required 9.6 °C for ripening

Fig. 39. Respiration curves of *Ulmus glabra* and *Picea abies* as a function of temperature. The respiration at 10 °C is taken as unity. After Skre (1983).

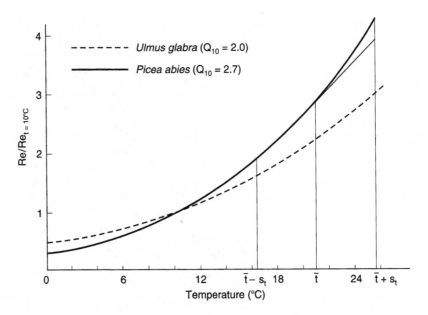

of the seeds. Such years are infrequent in subalpine areas of southern Norway and for this reason regeneration of spruce failed for many years in clear-felled areas.

It is not, however, only failure to produce ripe seeds that limits distribution. *Betula pubescens* ssp. *tortuosa* produces viable seeds every year near the timber-line, and *Ulmus glabra* produces seeds early in the growing season. There must be other factors operating.

To survive the winter the overwintering parts must ripen before the cold comes. As shown in experiments with frost-sensitive plants, frost resistance is highly dependent on the phenological phase (Sakai & Larcher 1987). A plant goes through a long series of processes from breaking dormancy in the spring, followed by the growth of shoots and stems, production of flowers and seeds, and finally the ripening of the seeds and buds before winter. All these processes require ATP. In cool summers the temperature may be insufficient for completing the preparations for the coming winter. If the overwintering parts are not fully ripened they might be killed in otherwise normal winters. For trees near the climatic timber-line the effect of cool summers appears as frost damage in the following winter, as shown by Wardle (1974) and Tranquillini (1979). In the Hamar district of Norway where there is a considerable production of apples, the summer of 1962 was unusually cold. During the following winter, which was not unusually cold, the trees were damaged

Fig. 40. Iversen diagrams of the thermospheres of *Picea abies* and *Ulmus glabra*. The isolines represent accumulated annual respiration (Re) at timber-line altitude. After Skre (1979b).

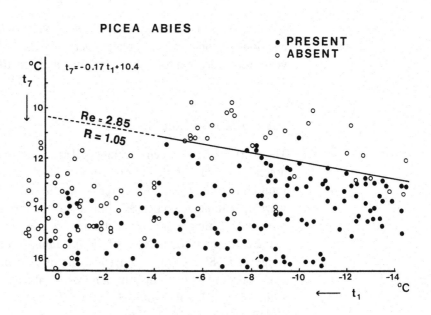

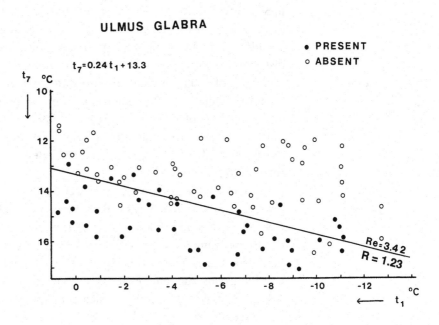

by frost. More than three-quarters of the apple trees were killed resulting in a serious setback for the apple growers. The apples which appeared next year were very malformed, suggesting damage to the buds (Thorsrud 1964).

Palaeoecological considerations

The correlations between R-values and species distribution can be used to assess earlier climatic conditions. Often fossil pollen, seeds and fruits are found north of and above the area of the present distribution, indicating a warmer summer climate. By comparison of former and present distributions an estimate of climatic differences can be made.

In Scandinavia fossil pollen of *Viscum album* is found beyond its present range limit. Skre (1979b) correlated its present distribution with a R-value of 7 and its fossil distribution with a value of 6. This corresponds to a difference in R-values of $1.95 - 1.79 = 0.16$. With a gradient of R of 0.135 per 100 m this corresponds to an altitudinal difference of 120 m or 0.85 °C. *Lycopus europaeus* today correlates with $R = 1.8$ and its fossil distribution in Scandinavia with 1.5. This corresponds to a 230 m altitudinal difference or a temperature difference of 1.6 °C. The present-day distribution of *Carex pseudocyperus* in Scandinavia correlates with 1.9 R whereas its fossil distribution correlates with 1.65. This corresponds to an altitudinal difference of 180 m or a temperature difference of 1.3 °C.

Moe & Odland (1992) estimated, based on pollen evidence, that the altitudinal limit of *Alnus incana* in western Norway was 175–200 m higher during the early and middle Holocene, corresponding to a temperature difference of about 1.7 °C. This is not corrected, however, for the isostatic recovery of the land.

Numerous subfossil stumps of birch and pine have been found above the present-day altitudinal limits of these species in the Norwegian mountains (Aas & Faarlund 1988). Compared with present-day values these occurrences amount to about 300 m. But as some isostatic recovery of the land has taken place, this must be accounted for. Based on such considerations Aas and Faarlund estimated that the temperature was 1.4–1.8 °C higher than today, which corresponds well with other results discussed above.

7   The boreal element

The boreal element in Europe is a northeastern element centred on the great conifer formation, the taiga, of northern Russia and Siberia. Boreal species form a series of equiformal progressive areas in the sense of Hultén (1937), from species with a very restricted distribution in northern Russia and perhaps penetrating into northern Finland to species with a much wider distribution, reaching the northern parts of the British Isles and the higher mountains of the Iberian Peninsula and the Balkans. It is typically a forest element. Only a few species extend north of the arctic forest-limit or much above the altitudinal timber-line.

The distribution patterns show that boreal species tend to be absent from areas with mild winters along the southwestern lowlands of Europe. Such species were called 'southwest coast avoiders' by Conolly & Dahl (1970) (see also Dahl 1951). They are able to tolerate high summer temperatures as shown by their occurrence in the lowlands of eastern and Central Europe. In this respect they differ from the true arctic-alpine plants. However, in the southwest they are restricted to higher elevations.

Climatic correlations

Many of the boreal species have distribution patterns that are the inverse of the atlantic species. For example, the horizontal distributions of *Picea abies* and *Ilex aquifolium* overlap only in the former Yugoslavia, but there *Picea* grows in the mountains whereas *Ilex* is restricted to lower elevations. Another example from Norway, pointed out by Blytt (1869), is that only in a restricted area along the Sognefjord does the atlantic species *Digitalis purpurea* occur in the same area as the boreal *Aconitum septentrionale*. Bøcher (1951) pointed out a similar inverse relation between the distribution of *Narthecium ossifragum* and the alpine *Rhododendron lapponicum*. Since the atlantic species are believed to be restricted to areas of high winter temperatures, it follows that boreal species are restricted to areas of cold winters and tend to be absent from areas with mild winters in the southwest.

One measure of winter severity is the mean temperature of the coldest month. It therefore seems reasonable to compare the distribution limits of boreal plants with isotherms of the coldest month. Since boreal species tend to occur at high elevations in the southwest, it is reasonable to compare their distributions with the mean temperatures of the coldest month calculated for the highest levels in the landscape, in a similar way as the winter conditions at the lowest stations were compared with the distribution of atlantic plants (see p. 40). The results are shown in Fig. 41, which is a generalised map, and in the following figures.

Based on information in *Atlas Florae Europaeae* and, for taxa not yet covered by this Atlas, on data in Hultén and Fries (1986) and other sources, the southern and western limits of many taxa have been found to correlate well with winter isotherms. The correlations are listed in Appendix II.

In general the southern and western distribution limits of many taxa compare well with the isotherms, with comparable limits in the different parts of Europe. However, there are some systematic discrepancies. Several taxa seem to tolerate warmer winters in Britain and, partly, in the

Fig. 41. Isotherms of the mean temperature of the coldest month of the year (°C) calculated for the highest points in the landscape.

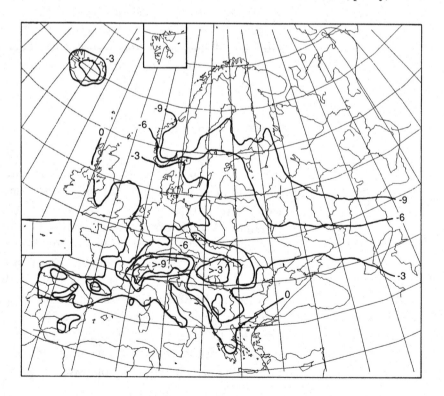

82

Low Countries than elsewhere. No explanation for this phenomenon can be given.

Within the boreal element the following series of sub-elements can be recognised.

1 *The Siberian boreal sub-element* reaching northern Russia and Finland and perhaps penetrating to northern Scandinavia along the Barents Sea. The limiting isotherms are from –6 to –12 °C. Examples include *Lycopodium complanatum* ssp. *montellii, Abies sibirica, Picea abies* ssp. *obovata* (Fig. 42), *Pinus sibirica, Salix jeniseiensis, Alnus viridis* ssp. *fruticosa, Cerastium regelii, Actaea erythrocarpa* and *Lactuca sibirica*. One species, *Delphinium elatum* (matching the –6 °C isotherm) is present also in the Alps (but missing in southwest Norway). However, there are some vicariants with a disjunct distribution between northern Russia and the Alps, for example *Abies alba* and *A. sibirica, Pinus cembra* and *P. sibirica*, and *Alnus viridis* ssp. *fruticosa* and ssp. *viridis*. The Russian distribution of members of this and the Scandinavian boreal sub-element has been discussed by Minyaev (1968).

2 *The Scandinavian boreal sub-element* reaching also the Baltic and the mountains of southwest Norway, but there restricted to higher altitudes. The limiting isotherms are between –3 and –9 °C. Examples include *Botrychium boreale* (Fig. 43), *Diplazium sibiricum, Aconitum septentrionale, Rubus arcticus, Epilobium hornemanni, E. lactiflorum, Petasites frigidus, Glyceria lithuanica, Carex disperma, C. tenuiflora* and *C. loliacea*. Some mosses also have a similar distribution, for example *Sphagnum wulfianum, Dicranum drummondii, Splachnum luteum* and *S. rubrum* (Størmer 1984).

3 *The North European boreal sub-element* reaching the Alps and/or the British Isles, but not extending to the Iberian Peninsula, Italy, or the Balkans. The limiting isotherms are from –2 to –6 °C. Examples include *Salix myrtilloides* (Fig. 44), *Stellaria calycantha* (Fig. 45), and *Betula nana* (Fig. 46). Other examples are *Botrychium virginianum, Salix phylicifolia, S. starkeana, Rubus chamaemorus, Polemonium coeruleum, Tofieldia pusilla* and *Poa remota*. Typically those species that occur in Scotland are found in the eastern, most continental parts of Scotland.

4 *The widespread boreal sub-element* reaching the higher mountains of the Iberian and Italian peninsulas and the Balkans. Their limiting isotherms are from –2 to +3 °C. Examples are *Picea abies* (Fig. 47), *Pinus*

Fig. 42. Distribution of *Picea abies* spp. *obovata* in relation to the –10 °C winter temperature isotherm. Filled circles show species presence, open circles the other grid squares of *Atlas Florae Europaeae* where the coldest month at the highest point has –10 °C or less. Points show presence in grid squares outside the indicated isotherm.

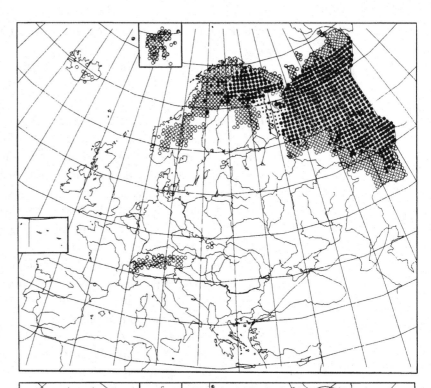

Fig. 43. Distribution of *Botrychium boreale* in relation to the –5 °C winter temperature isotherm. Filled circles show species presence, open circles the other grid squares of *Atlas Florae Europaeae* where the coldest month at the highest point has –5 °C or less. Points show presence in grid squares outside the indicated isotherm.

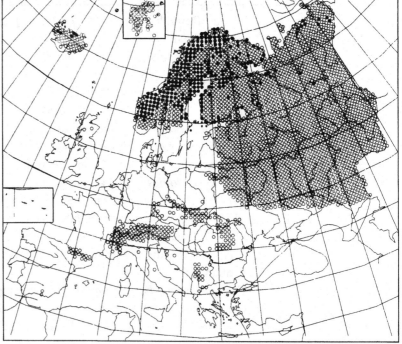

Fig. 44. Distribution of *Salix myrtilloides* in relation to the –4 °C winter temperature isotherm. Filled circles show species presence, open circles the other grid squares of *Atlas Florae Europaeae* where the coldest month at the highest point has –4 °C or less. Points show presence in grid squares outside the indicated isotherm.

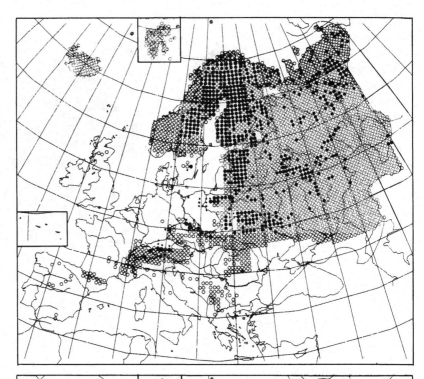

Fig. 45. Distribution of *Stellaria calycantha* in relation to the –3 °C winter temperature isotherm. Filled circles show species presence, open circles the other grid squares of *Atlas Florae Europaeae* where the coldest month at the highest point has –3 °C or less. Points show presence in grid squares outside the indicated isotherm.

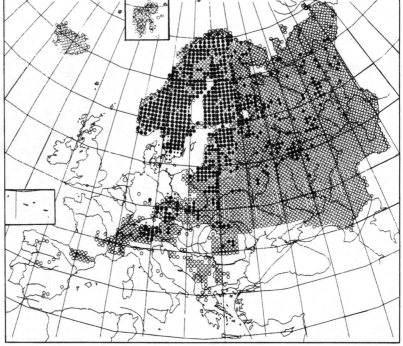

85

Fig. 46. Distribution of *Betula nana* in relation to the −2 °C winter temperature isotherm. Filled circles show species presence, open circles the other grid squares of *Atlas Florae Europaeae* where the coldest month at the highest point has −2 °C or less. Points show presence in grid squares outside the indicated isotherm.

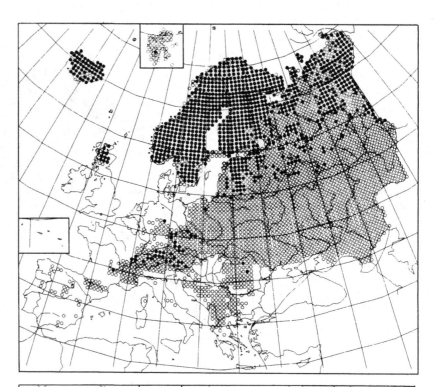

Fig. 47. Distribution of *Picea abies* in relation to the −2 °C winter temperature isotherm. Filled circles show species presence, open circles the other grid squares of *Atlas Florae Europaeae* where the coldest month at the highest point has −2 °C or less. Points show presence in grid squares outside the indicated isotherm.

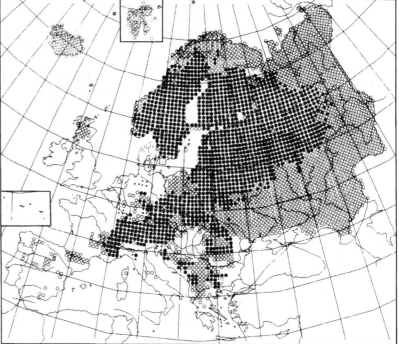

sylvestris (Fig. 48), *Lychnis viscaria* (Fig. 49) and *Gymnocarpium dryopteris* (Fig. 50). Other examples include *Lycopodium annotinum, Thelypteris phegopteris, Salix pentandra, S. caprea, Alnus incana, Silene rupestris, Ranunculus platanifolius, Geranium sylvaticum, Viola mirabilis, Pyrola* spp., *Vaccinium* spp., *Trientalis europaea, Galium boreale, Melampyrum arvense* and *M. sylvaticum, Maianthemum bifolium, Melica nutans, Carex digitata* and *Listera cordata.* Many of these species are typical of the boreal conifer forests. Many bryophytes of the same communities also have a similar distribution pattern, for example *Dicranum majus, Pleurozia schreberi* and *Hylocomium splendens.*

A rather peculiar case is *Woodsia alpina,* whose distribution on the continent correlates with the −6°C isotherm, but in Scotland with the −3°C isotherm and in Wales with the 0°C isotherm.

In these elements there are certain groupings of species according to their life forms. There is a high proportion of woody species (trees, shrubs and ericoids). There are few annuals; *Sedum annuum, Saxifraga adscendens* and *Androsace septentrionalis* are the only annuals or bi-ennials. Very few are aquatic plants. If there is one boreal species in a genus, there tend to be several. Genera with many such boreal species include *Botrychium, Salix, Stellaria, Corydalis, Alchemilla, Eriophorum* and *Carex.* There are few boreal members of the Poaceae and Fabaceae.

Several lichens and bryophytes have similar distribution patterns (Ahti 1977; Hill *et al.* 1991). Many of the typical spruce forest liverworts exhibit a boreal distribution pattern, for example *Barbilophozia floerkei, B. hatcheri, B. lycopodioides* and *Ptilidium ciliare* (Hill *et al.* 1991). Among the mosses there are *Ptilium crista-castrensis* and *Hylocomium splendens* and, among the lichens, *Cetraria delisei.*

Ecophysiology

As emphasised earlier, a correlation is not an explanation, it is a description. In order to be meaningful a physiological mechanism that can be tested by experiment is needed to account for the correlation. However, at present only tentative suggestions can be given for the boreal element. This is a research area requiring further attention.

One possible mechanism has been suggested by Mork (1938) (see also Robak 1960). In plantations in western Norway, outside its natural range,

Fig. 48. Distribution of *Pinus sylvestris* in relation to the –1 °C isotherm for the coldest month and the +33 °C maximum summer temperature. Filled circles show species presence, open circles the other grid squares of *Atlas Florae Europaeae* where the coldest month at the highest point has –1 °C or less and maximum summer temperatures not higher than +33 °C. Points show presence in grid squares outside the indicated isotherm.

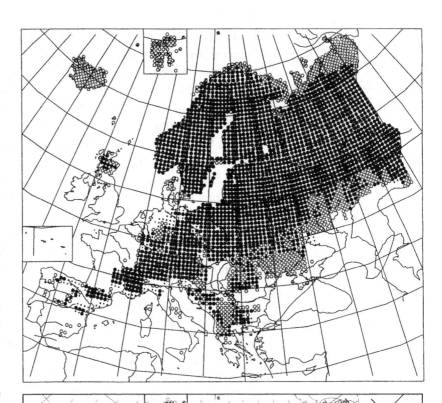

Fig. 49. Distribution of *Lychnis viscaria* in relation to the +1 °C winter temperature isotherm. Filled circles show species presence, open circles the other grid squares of *Atlas Florae Europaeae* where the coldest month at the highest point has +1 °C or less. Points show presence in grid squares outside the indicated isotherm.

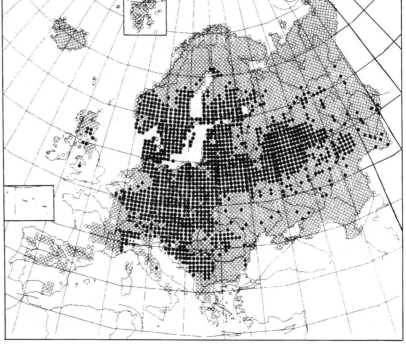

Fig. 50. Distribution of *Gymnocarpium dryopteris* in relation to the +1 °C winter temperature isotherm. Filled circles show species presence, open circles the other grid squares of *Atlas Florae Europaeae* where the coldest month at the highest point has +1 °C or less. Points show presence in grid squares outside the indicated isotherm.

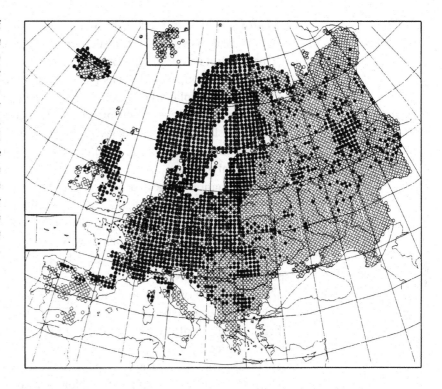

Picea abies can grow well and can often produce more timber than in forests farther east due to the abundant moisture and a longer growing season. It also produces abundant germinable seeds. But such forests do not reproduce effectively by natural regeneration, and after felling a new forest stand must be planted. This suggests that the limiting factor may be related to the establishment of the offspring.

If seeds of spruce are taken into the laboratory and wetted and dried at temperatures too low for germination, they rapidly lose their viability. Seeds of *P. abies* are shed in the autumn and lie on the ground during winter. If they are covered by snow, the seeds remain viable, but without a snow cover they die. Many boreal species have limits that correlate with temperatures at around or a little below 0 °C which is at the limit of a consistent snow-cover in winter.

Pinus sylvestris, which is well established along the coasts of Norway and Scotland (Fig. 48), has seeds that lose their germination ability if treated in the same way as spruce seeds. However, pine seeds are shed later in winter than the seeds of spruce, and remain protected in the cones sufficiently long to survive the critical period. Sitka spruce (*Picea sitchensis*), which is often able to reproduce in plantings in western Norway, sheds its seeds at a time between pine and spruce.

This may not be the whole story concerning the limited success of *Picea abies* in coastal climates. If it was the full explanation, spruce would be a successful tree in plantations over a considerable area outside its present natural range. However, such plantings in England have generally not been successful and other species are now planted instead. While this is an explanation pertaining to conifers, many angiosperms have similar geographical patterns.

An explanation suggested by Printz (1933) for spruce is that during mild winters with little light, the respiration rate is so high that carbohydrate reserves are used up. This does not fit well with the high production in spruce plantations in western Norway outside its natural limit (Hagem 1947). It might, however, be an explanation for other boreal species.

An experience of many gardeners is that it is often necessary to cover plants in winter and spring to prevent them starting their growth too early. There is insufficient winter stability, and this is a plausible hypothesis for the avoidance by many plants of areas with mild winters. To test this hypothesis a number of boreal plants were grown in an experimental garden at Stavanger in southwestern Norway, outside the natural range of the species. The results are reported by Salvesen (1988, 1989). In general the boreal species, or boreal populations of more widespread species, had a tendency to continue growth during autumn and winter and were damaged by frost. Frequently they survived to the next summer, but after a few years they died. The southwestern plants tended also to flower later in summer than the boreal plants. An interesting example is the species pair *Omalotheca sylvatica*, which is not a boreal species, and the boreal *O. norvegica*. The phenological cycle of *O. sylvatica* is controlled by daylength and the plant remains dormant until the danger of late frost is over. In autumn it goes into dormancy as the days shorten. *Omalotheca norvegica* has no daylength control of phenology and grows during mild periods in midwinter in Stavanger.

It is known that the responses to daylength in many species are adaptable and that different strains of one species can differ in such responses. Why do some species only have a limited ability to develop strains adapted to mild winters, while others do not? This is a field of research which has not been systematically explored and several mechanisms are possible.

Palaeoecological implications

From the correlations between winter temperatures and geographical distribution, the occurrence of boreal species outside their present range limits could be ascribed to changes in winter climate. Thus *Betula nana* has been found to have occurred in numerous places outside its present range during glacial and late-glacial times (Conolly in Conolly & Dahl 1970; Hultén & Fries 1986) which might thus be ascribed to colder winters in the past. *Betula nana* has a limiting winter isotherm of −2 °C but its fossils are present in late-glacial deposits in Cornwall with present-day winter temperatures of +3 °C, a difference of 5 °C. In southwest France *Pinus sylvestris* occurred at 40 m altitude during the Weichselian glacial maximum, while today it is restricted to altitudes above 1500 m (Oldfield 1964). Since *Pinus sylvestris* is a boreal species avoiding areas with mild winters this implies a temperature depression of at least 10 °C in the winter months. A winter temperature depression of the order of 20 °C accords with the former presence of *Pedicularis lanata*, *P. hirsuta* and *Silene furcata* (= *Lychnis furcata*) in Britain (Godwin 1975; Dickson 1992) and of several species of beetles in England not known nearer than northern Russia today (Coope *et al.* 1971). The climatic simulation model reconstruction by Kutzbach & Guetter (1986) of January temperatures 18 000 years ago suggests temperature differences from present of the order of 20 °C (see p. 33).

According to the climatic interpretation of fossils of atlantic plants outside their present area, the winters during the post-glacial thermal maximum were about 4 °C warmer than today. The distribution of spruce at that time should accordingly be expected to correlate with the −6 °C winter temperature isotherm. Figure 51 shows the isopolls of spruce according to Huntley & Birks (1983) compared with the −6 °C winter isotherm. There is a gap between a potential spruce area in the Central European mountains and a spruce area in northern Russia and Scandinavia. When winter temperatures became sufficiently cold, spruce could migrate from Central Europe to the Baltic countries and interbreed with the previous spruce population, which may have been ssp. *obovata*. Subsequent spread to the west and south occurred, and spruce arrived in Norway only about 2000 years ago (Hafsten 1991).

Fig. 51. Isopoll map of the distribution of *Picea abies* 5000 years ago (after Huntley & Birks 1983) compared with the –6 °C winter temperature isotherm. Filled circles show the presence of spruce pollen inside the area delimited by the –6 °C isotherm; the solid line represents presence of spruce pollen outside this area at the 5% level, and the broken line at the 2% level.

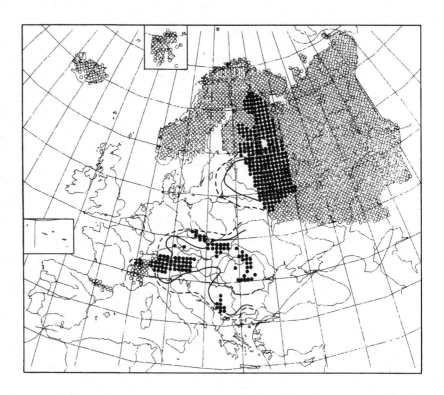

8 The arctic, alpine and montane elements

Arctic plants have their main distribution north of the arctic timber-line. Alpine plants have their main distribution above the climatic alpine timber-line. Montane plants have their main distribution in high-lying forests in the southern part of their area but can occur at low elevations in the north. Their lower distribution limits fall from the south towards the north and generally from continental to oceanic areas. Many otherwise alpine species can be found at sea level along the west coasts of the British Isles and Fennoscandia.

The alpine and montane plants have a distribution approximately the inverse of the thermophilic plants. Since the upper and northern limits of the thermophilic plants are related to summer temperature, the lower and southern limits of the alpine plants must be negatively related to high summer temperatures. The question is, what are the physiological mechanisms responsible for these distribution patterns.

One obvious explanation is competition (Pigott 1978). Alpine and arctic plants are low-growing and hence unable to compete for light where climate is favourable for tree growth. This hypothesis can easily be tested by planting alpine species in botanic gardens in the absence of competition and observing whether they survive. Often they do quite well, but not always. Some species are hard to keep outdoors for many years in botanic gardens. Very often they survive the spring but seem to suffer in the summer heat. They have a better period in autumn and next spring, but after a few years they are so weakened that they die. This is the case with, for instance, *Salix herbacea* or Norwegian populations of *Oxyria digyna* in the Botanic Garden in Oslo. In Copenhagen an arctic greenhouse was built with special cooling in summer which enabled Danish botanists to grow successfully plants from Greenland that they could not grow outdoors. It is possible to grow some high-arctic plants under controlled conditions in a phytotron.

Some alpine species are tolerant of shade, for example those that are able to compete and survive in tall-herb meadows or in dense willow scrub. In cultivation they do better in half-shade than in direct sunlight.

Climatic correlations

Dahl (1951) suggested that some factor associated with maximum summer temperatures was critically limiting the performance of alpine and arctic species. To test this hypothesis mean annual maximum summer temperatures from meteorological observations were calculated, that is the mean of the maximum temperatures measured each year for a 15 year period. Then the expected maximum summer temperatures at the highest summits in the landscape were calculated, assuming a vertical temperature gradient of 0.6 °C per 100 m altitudinal difference. Isotherms separating, for instance, an area with summits cooler than 26 °C from areas where no such localities existed, were drawn and finally these isotherms were compared with the distribution maps of Fennoscandian plants just published by Hultén (1950). An example is given in Fig. 52. Similar comparisons have been made for the British Isles by Conolly & Dahl (1970).

For comparison with the distribution limits towards the south and towards the lowlands, maximum summer temperatures have been calculated for the entire *Atlas Florae Europaeae* area. The calculations are explained in Appendix I. The mean of the annual maxima observed at the stations was taken, and the gradient used was 0.7 °C per 100 m. A generalised map of this parameter for Europe is given in Fig. 53. It must be pointed out that the meteorological data from northern Russia are based on only a few years' observations. Moreover, few stations are there, and none near the Arctic Ocean with its chilling effect on maximum temperatures. Thus any correlations in this area are probably not reliable.

Based on comparisons between the maximum summer temperatures and the distribution of plant species the following elements are recognised.

1 **The arctic element** typical of the area north of Fennoscandia or penetrating into the area at high altitudes in northernmost Fennoscandia. Arctic species are generally distributed in Siberia, Novaya Zemlya, Svalbard, northern Greenland, and Arctic Canada and exhibit a Circumpolar or Amphi-Atlantic distribution pattern. A list of such species is given in Appendix III. Typical examples are *Dupontia* spp., *Pleuropogon sabinei*, *Poa abbreviata*, *Eriophorum triste*, *Saxifraga flagellaris* and *Pedicularis lanata* coll. More than half of them are graminoids. Among the lichens *Neuropogon sulphureus* and *Dactylina arctica* may be mentioned.

Fig. 52. The distribution of *Salix herbacea* in relation to the 26 °C maximum summer temperature isotherm (after Dahl 1951). Hatched area represents frequent occurrences; filled circles show more isolated localities; open circles show fossil occurrences in south Scandinavia. (Crosses represent fossil finds of *S. polaris*).

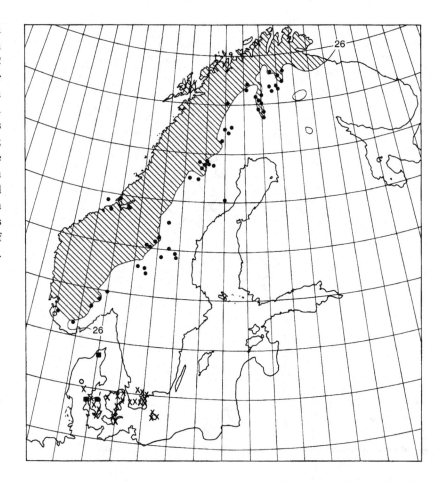

2 **The alpine element** occurring in the regions of the mountains, many also in the high arctic. Within this element there are different sub-elements.

   (a) *A high-alpine sub-element*. A typical example is *Ranunculus glacialis* (Fig. 54). The members of this group have limiting maximum summer isotherms of +22 °C or lower. Other examples are *Sagina nivalis*, *S. caespitosa*, *Cassiope hypnoides*, *C. tetragona*, *Campanula uniflora*, *Erigeron uniflorus*, *Luzula arctica*, *Poa flexuosa*, *Carex misandra* and *C. rufina*. Most of them occur in Iceland and only a few occur in the British Isles. Some of this group have been found as late-glacial fossils in Scotland, for instance a scapiflorous *Papaver* that subsequently died out, probably when the climate became too warm (Conolly & Dahl 1970). Many of them have a so-called bicentric distribution in Scandinavia as they are

Fig. 53. Isotherms of the mean annual maximum summer temperature calculated for the highest points in the landscape.

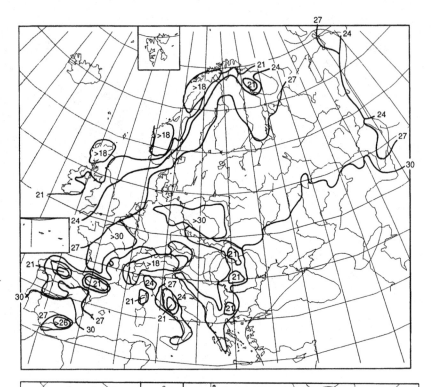

Fig. 54. Distribution of *Ranunculus glacialis* in relation to the +22 °C isotherm of maximum summer temperature. Filled circles show species presence, open circles the other grid squares of *Atlas Florae Europaeae* where the warmest month at the highest point has +23 °C or less. Points show presence in grid squares outside the indicated isotherm.

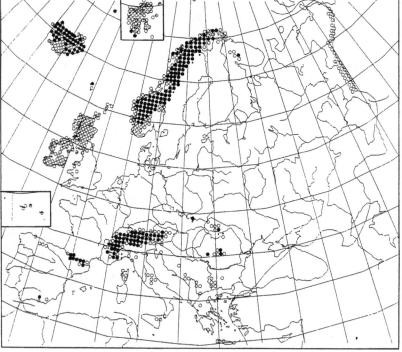

missing from the areas of lower mountains between the Dovre mountains in southern Norway and the high mountains in northern Norway. Many have an Amphi-Atlantic distribution.

(b) *A mid-alpine sub-element* limited by maximum summer isotherms of +25 °C or lower. Typical examples are *Cerastium arcticum* coll. (Fig. 55), *Salix herbacea* (Fig. 56) and *Silene acaulis* (Fig. 57). Other examples include *Salix reticulata, Koenigia islandica, Sedum rosea, Saxifraga rivularis, Diapensia lapponica, Luzula spicata* and *Carex saxatilis.* All of these examples occur in the British Isles.

(c) *A low-alpine sub-element* limited by maximum summer temperatures of +27 °C or lower, that also extends down into subalpine areas. Examples are *Cerastium cerastoides* (Fig. 58) and *Diphasium alpinum* (Fig. 59). Other members of this group include *Cryptogramma crispa, Athyrium distentifolium, Oxyria digyna, Sagina saginoides, Thalictrum alpinum, Alchemilla alpina, Epilobium alsinifolium, Viola biflora, Omalotheca supina, O. norvegica, Carex bigelowii, C. norvegica* and *C. lachenalii.* Many of these species occur in the British Isles.

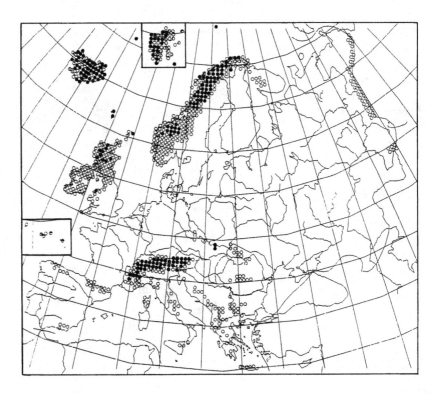

Fig. 55. Distribution of *Cerastium arcticum* (including *C. uniflorum* in the Alps) in relation to the +23 °C maximum summer temperature isotherm. Filled circles show species presence, open circles the other grid squares of *Atlas Florae Europaeae* where the warmest month at the highest point has +23 °C or less. Points show presence in grid squares outside the indicated isotherm.

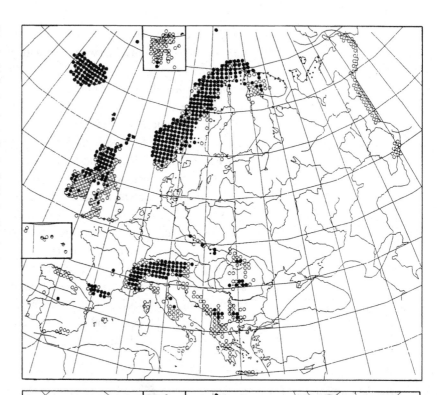

Fig. 56. Distribution of *Salix herbacea* in relation to the +24 °C maximum summer temperature isotherm. Filled circles show species presence, open circles the other grid squares of *Atlas Florae Europaeae* where the warmest month at the highest point has +24 °C or less. Points show presence in grid squares outside the indicated isotherm.

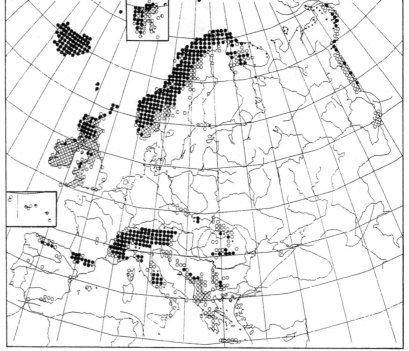

Fig. 57. Distribution of *Silene acaulis* in relation to the +24 °C maximum summer temperature isotherm. Filled circles show species presence, open circles the other grid squares of *Atlas Florae Europaeae* where the warmest month at the highest point has +24 °C or less. Points show presence in grid squares outside the indicated isotherm.

Fig. 58. Distribution of *Cerastium cerastioides* in relation to the +26 °C maximum summer temperature isotherm. Filled circles show species presence, open circles the other grid squares of *Atlas Florae Europaeae* where the warmest month at the highest point has +24 °C or less. Points show presence in grid squares outside the indicated isotherm.

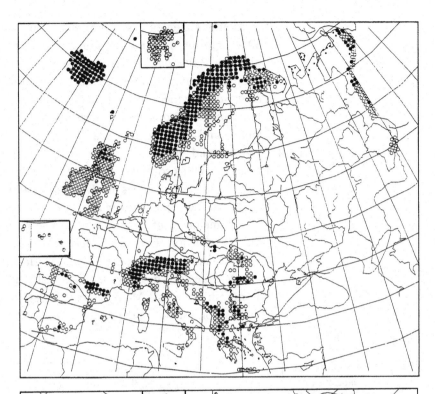

Fig. 59. Distribution of *Diphasium alpinum* in relation to the +27 °C maximum summer temperature isotherm. Filled circles show species presence, open circles the other grid squares of *Atlas Florae Europaeae* where the warmest month at the highest point has +27 °C or less. Points show presence in grid squares outside the indicated isotherm.

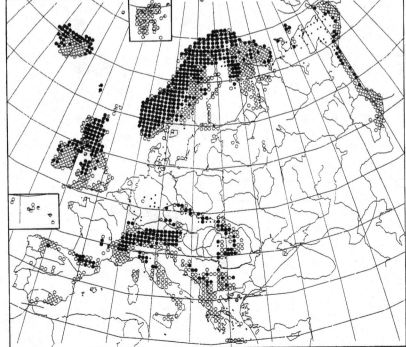

3 **The montane element** limited by maximum summer temperatures of +30 °C or lower. Examples are *Rumex alpestris* (Fig. 60) and *Pinus mugo* (including *P. uncinata*) (Fig. 61). Other members of this element include *Dryopteris assimilis, Pinus cembra, Salix phylicifolia, Betula nana, Polygonum viviparum, Alchemilla glomerulans, Potentilla crantzii, Empetrum hermaphroditum, Bartsia alpina, Saussurea alpina, Phleum alpinum* and *Carex magellanica*.

4 **A widespread element** avoiding the the warmest parts of southern Russia as well as southwestern France and with limiting isotherms of +33 °C or lower. Examples are *Salix lapponum* (including *S. helvetica*) (Fig. 62) and *Gymnocarpium dryopteris* (Fig. 63, cf. Fig. 50).

Other members of this widespread element are species typical of boreal conifer forests such as *Picea abies* ssp. *abies, Pinus sylvestris, Lycopodium annotinum, L. clavatum, Maianthemum bifolium* and *Linnaea borealis*. They have a boreal distribution pattern in the west, but not in the east. Many other species have similar distribution patterns, for example *Stellaria longifolia, Prunus padus, Alchemilla*

Fig. 60. Distribution of *Rumex alpestris* in relation to the +28 °C maximum summer temperature isotherm (in the Alps ssp. *alpestris*, in the north ssp. *lapponica*). Filled circles show species presence, open circles the other grid squares of *Atlas Florae Europaeae* where the warmest month at the highest point has +28 °C or less. Points show presence in grid squares outside the indicated isotherm.

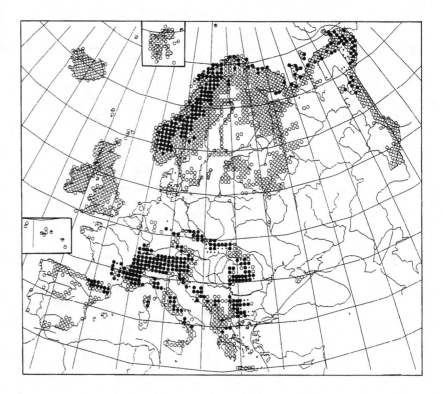

100

Fig. 61. Distribution of *Pinus mugo* (including *P. uncinata*) in relation to the +30 °C maximum summer temperature isotherm. Filled circles show species presence, open circles the other grid squares of *Atlas Florae Europaeae* where the warmest month at the highest point has +30 °C or less. Points show presence in grid squares outside the indicated isotherm.

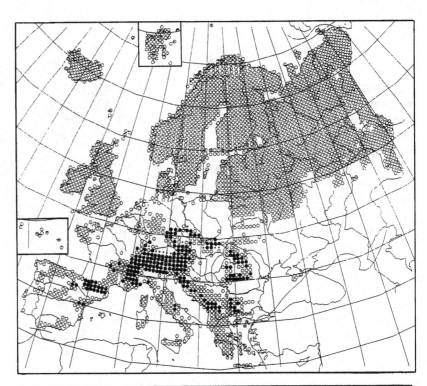

Fig. 62. Distribution of *Salix lapponum* (including *S. helvetica*) compared with the +31 °C maximum summer temperature isotherm and the –3 °C isotherm of the coldest month. Filled circles show species presence, open circles the other grid squares of *Atlas Florae Europaeae* where the warmest month at the highest point has +31 °C or less and the coldest month has –3 ° or warmer. Points show presence in grid squares outside the indicated isotherm.

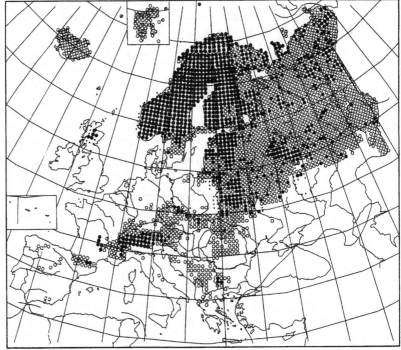

101

Fig. 63. Distribution of *Gymnocarpium dryopteris* compared with the +32 °C maximum summer temperature isotherm. Filled circles show species presence, open circles the other grid squares of *Atlas Florae Europaeae* where the warmest month at the highest point has +32 °C or less. Points show presence in grid squares outside the indicated isotherm.

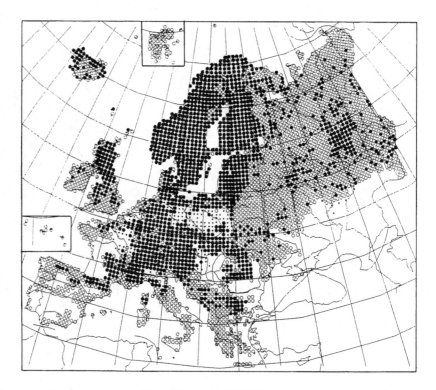

glabra, Geranium sylvaticum, Cirsium helenioides, Nardus stricta and *Eriophorum vaginatum*. The element thus partly overlaps with the widespread boreal sub-element (p. 83).

Ecophysiology

One hypothesis to explain the observed correlations is that the plants die from overheating. Resistance to overheating can be measured by exposing plants to high temperatures and subsequent cultivation to observe whether they survive. A suitable method is to put plant parts into a hot water-bath for given periods of time, for instance to have three or four water-baths at 3–4 °C intervals between +40 and +60 °C (see Fig. 64) and to place plants or plant parts in the water for limited periods, for example 2 hours, 1 hour, 30 minutes, 15 minutes, 7.5 minutes, and 3.75 minutes. By subsequent cultivation their survival can be observed. This can be done visually or, better, by electrometric methods.

Figure 64 gives the results for *Astragalus norvegicus* and *Salix reticulata* where Arrhenius plots are used, with the logarithm of time of insertion

Fig. 64. Lethal temperatures for (*a*) *Astragalus norvegicus* (+47.5 °C) and (*b*) *Salix reticulata* (+44.7 °C). Values for duration of insertion of plant parts in hot water and water temperatures are indicated in Arrhenius plots, with the logarithm of time along the vertical axis and the inverse of absolute temperature along the horizontal axis. Open circles mean that plants survived, crosses that they died, circle with cross that 50% of the leaves were still alive. (*a*) after Gauslaa (1984), (*b*) after Kjelvik (1976).

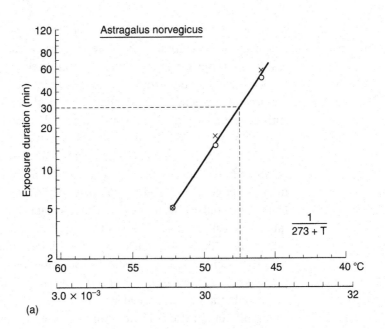

(a)

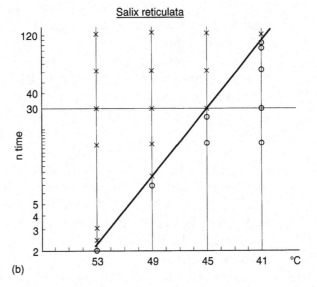

(b)

along the y-axis and the inverse of absolute temperature (°K) measured in the water-bath along the x-axis. Open circles indicate that the plants survived, crosses that they died. It is seen that a straight line separates the circles and the crosses. This enables the calculation of an activation energy of the reaction leading to the death of the plants of 250–560 kJ/mol. Such values are typical for denaturation reactions of proteins as a function of temperature. It is possible that the heat causes an inactivation of some essential enzyme leading to death. The intersection of the curve with the vertical axis at 30 minutes is defined as the lethal temperature. The lethal temperature depends on the phenological phase and is lowest in early spring and highest in winter. Hardening at elevated temperatures for a day or so also increases the lethal temperature.

Lethal temperatures have been measured for numerous plant species under different environmental conditions. The range of lethal temperatures is between +42 and +56 °C irrespective of whether the plants are from subtropical forests, deserts, Mediterranean scrub, or alpine vegetation (Gauslaa 1984). The limiting isotherms of the North European plants are in the range of +20 to +34 °C, that is some 20° lower than the lethal temperature. The question arises whether it is possible that the plants as they grow in nature can be heated 20 °C above the temperature measured in a meteorological screen.

A good example for consideration is the small annual *Koenigia islandica* (Dahl 1963a). It grows in open habitats exposed to the sun and in permanently wet sites. It can be thought of as growing on a wet surface.

The surface is heated by the solar radiation, S, of which a fraction, S_j, is absorbed. Heat is exported to the surroundings by heating of the air (H_c), by evaporation of water (H_v), by heat radiation to the atmosphere (R_s), minus heat radiation from the atmosphere (R_a), and finally by heating of the soil (H_s). Conservation of energy gives the following equation:

$$S_j = H_c + H_v + R_s - R_a + H_s.$$

In Fig. 65 the different components of heat loss from a surface with +25 °C air temperature and 50% relative humidity are given as a function of surface temperature. The important components of heat loss are by evaporation of water, by net heat radiation, and by heating of the soil. The loss of heat to the atmosphere and to space is shown as well as the total heat loss.

Solar radiation on a clear day in Northern Europe in summer is about 5.4 J/cm² min (= 1.3 cal/cm²) of which 85% is absorbed, that is 4.6 J/cm² min (= 1.1 cal/cm²). From Fig. 65 it is seen that the total outward loss of heat is about 1.1 cal/cm² at +53 °C so that this temperature cannot be reached. The total loss of heat including heating of the soil is 1.1 cal/cm² at +43 °C. Thus the probable maximum temperature experienced by *Koenigia* is between +43 and +53 °C. The limiting maximum summer

Fig. 65. Components of heat loss of a wet vegetation surface as a function of surface temperature when atmospheric temperature is +25°C and relative humidity 50% (Dahl 1963a). Surface temperature along the horizontal axis. Further explanation is given in the text.

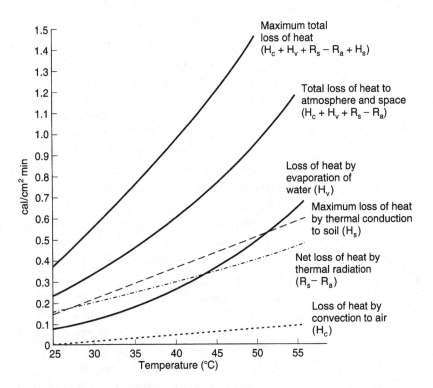

isotherm of the species is +23 °C while its lethal temperature is +44 °C (Gauslaa 1984). Thus it is quite possible that the distribution of *K. islandica* is directly limited by its sensitivity to overheating. Gauslaa found, by comparing lethal temperatures of cushion species like *Silene acaulis* with temperatures observed during hot days, that they could also be limited directly by their sensitivity to overheating.

The temperature experienced by plants also depends on the content of water vapour in the atmosphere. High humidity reduces loss of heat by evaporation and heat radiation from the atmosphere increases with humidity. Thus at the same air temperature plants are heated more in an oceanic than in a continental climate. Comparison of a site for *Koenigia* with 50% relative humidity with a site with 10% relative humidity showed that *Koenigia* is expected to become 3.2 °C hotter in the moist than in the dry atmosphere. Interestingly, Conolly & Dahl (1970) found that the limiting isotherms of alpine and montane plants were, on an average, 1.64 °C lower in northern England and southern Scotland than in continental parts of Scandinavia.

In Colorado the lowermost localities of *Koenigia* have an annual maximum summer temperature of +19 °C, that is 5° below the value for Fennoscandia. However, at higher altitudes and lower latitudes incoming solar radiation is higher, and other components in the calculations are also affected. These differences are calculated to amount to a temperature difference of about 5 °C, close to the observed difference.

These calculations apply to plants that grow at or close to the soil surface. A difference of 20 °C between the surface of a grass turf well supplied with water and the atmospheric temperature measured at 2 m altitude on a clear, still summer day is considered as normal and corresponds to calculations based on free convection (Sutton 1953; Gates 1980). Other conditions apply to more erect plants, where both morphology and diffusion resistances affect the temperature of the plants.

A classical example has been provided by Lange (1959). He measured plant temperatures and lethal temperatures in the deserts of Mauritania. He found one group of plants whose leaf temperatures were consistently lower than atmospheric temperatures, due to the expenditure of heat by evaporation of water. Another group had higher leaf temperatures than air temperatures. The lethal temperatures of the species in the first group were consistently lower than those in the second group. He then deprived plants in the first group of access to water by cutting their stems. Leaf temperatures then rose high above the lethal temperature of the species (Fig. 66). These plant species thus clearly depend upon a supply of water to avoid overheating.

Fig. 66. Leaf temperatures in *Citrullus colocynthis* (L) compared with air temperatures (A) during 7 July 1956 in Mauritania. At 12.30 p.m. a stem of *Citrullus* was cut and its temperature (B) rapidly exceeded its lethal temperature of +51 °C. After Lange (1959).

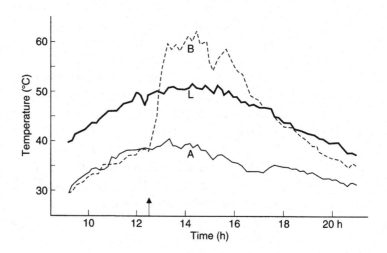

The cooling of plants depends upon the diffusion resistances in the leaves. Some species have a low resistance and transpire nearly as a wet surface. Others have a high diffusion resistance and conserve water by closing their stomata. Gauslaa (1984) has measured such resistances in many Nordic species. *Sedum rosea, S. telephium, Polygonatum odoratum, Diapensia lapponica* and, in general, dry-habitat species have a high cuticular resistance and conserve water by closing their stomata. On the other hand, *Pedicularis oederi, Saxifraga aizoides, Oxyria digyna* and, in general, moist habitat species show low cuticular diffusion resistances and depend on a steady supply of water. In general wet-habitat plants within the alpine flora have consistently lower lethal temperatures than the dry-habitat plants.

Leaf temperatures of the plants also depend on the external morphology of the plants. The heat-exchange properties depend upon the characteristic dimension, which for a disc is the diameter of the disc, for cylinders the diameter of the cylinder. Thus plants with small or thin leaves or with thin, cylindrical shoots are efficient heat exchangers. Broad-leaved plants and also cushion plants are inefficient heat exchangers and are heated by solar radiation.

It is typical of the vegetation of hot and dry environments that the plants are built as efficient heat exchangers. An exception is the succulents, which can stop transpiration by closing their stomata. However, they have high lethal temperatures and a special biochemistry to utilise the carbon dioxide produced by respiration at night so that it is later used for photosynthesis during the day. The non-succulents have finely dissected or compound leaves as in the Fabaceae, narrow and thin leaves as in tussock grasses, or thin, cylindrical, assimilating stems as in the *palo verde* formations. Such forms can be seen as an adaptation to avoid overheating of the plants, by dispersing the heat absorbed by the plant to the environment as sensible heat without an unnecessary expenditure of water (Dahl 1966). It is in habitats with a good supply of water that plants with broad and large leaves grow, for example in tropical rain forests, temperate deciduous forests, or in tall-herb meadows. Plants with broad leaves run the risk of overheating if water is limited.

Different physiological mechanisms can result in similar climatic correlations. They must be analysed in each separate case. For instance, the alpine species *Loiseleuria procumbens* (limiting isotherm +25 °C, lethal temperature +51.2 °C) and *Saxifraga aizoides* (limiting isotherm +26 °C, lethal temperature +46.3 °C) have similar distribution patterns but different ecology. According to Gauslaa (1984, p. 72) it seems unlikely that *S. aizoides* with a high transpiration rate can be heated above its lethal temperature in the mountains, while this is quite possible for *L. procumbens* with a low transpirational heat loss. In the western Alps *S. aizoides* can be found along calcareous seepages at low elevations. What physiological mechanism limits the geographical distribution of *S. aizoides* is unknown.

Russian scientists (Alexandrov 1977; Denko *et al.* 1981) doubt that heat resistance is of adaptive significance in plants, despite all the correlations between morphology and diffusion resistances. They believe that low heat resistance reflects the conformational flexibility of the molecules within the plant cells that enable them to function well at low temperatures. If so, one would expect that the lethal temperatures of alpine and arctic plants would differ from those of more temperate species. However, by comparing the ranges of lethal temperatures in the flora of such climatically different places as Puerto Rico (Biebl 1965), woodlands and savannas of East Australia (E.D., unpublished results), the Mauritanian desert (Lange 1959), and Costa Brava in Spain (Lange & Lange 1963), with those of alpine species in Norway (Gauslaa 1984, p. 28) and Greenland (Biebl 1965), no significant differences appear. Denko *et al.* (1981) found less than 2 °C difference in thermostability between arctic plants in Franz Josefs Land and Siberia and boreal plants near St Petersburg.

Many arctic and alpine plants are built as inefficient heat exchangers and keep their temperatures high for sufficient ATP production (Dahl 1986 and above; Gauslaa 1984). But the adaptations for heat conservation at the same time make the plants vulnerable to overheating in warmer climates. This point can be illustrated by the ecology of *Koenigia islandica* in Svalbard where it grows close to the ground, hardly more than 2 mm high. Its lethal temperature is +44 °C and Gauslaa (personal communication) found that the seeds did not germinate well at temperatures below +20 °C. These cardinal temperatures are not very different from what might be found for many forest trees. However, conditions satisfying the requirements of *Koenigia* are in Svalbard only realised in a thin layer above and below the soil surface.

Obviously, different physiological mechanisms can result in similar distribution patterns and it may be difficult from geographical data alone to decide what are the important causal factors. Within a limited geographical area it may be especially difficult to decide whether the distribution of a species is limited by a need for cold winter temperatures or for low summer temperatures, since both restrict the distribution to high mountains. It is only by detailed autecological studies that such cases can be decided.

Palaeoecological considerations

More than 100 years ago Nathorst (1892) found subfossil leaves of alpine species like *Salix herbacea* in late-glacial deposits in southernmost Scandinavia, showing that the climate there had once been favourable for alpine plants, and that they had the opportunity to migrate from the south to the mountains. Later numerous finds have been made of many species both in Fennoscandia and in the British Isles (Godwin 1975).

Conolly (in Conolly & Dahl 1970) found that she could account for all subfossil finds of alpine-montane plants in Britain by assuming a depression of maximum summer temperatures of 6°C during the Weichselian maximum. This is corroborated by a comparison of the present and subfossil occurrences of *Salix herbacea* in Europe where the subfossil finds are limited by the +30°C isotherm whereas the present-day distribution is delimited by the +24°C isotherm. For the Younger Dryas stadial she found a depression of 3°C, similar to what Dahl (1964) had found for southern Scandinavia. Similar comparisons in eastern North America suggest a depression of 4°C for the maximum Wisconsin glaciation (Dahl 1963a). These estimates compare well with the results of climatic reconstructions of Ice Age conditions.

9 Endemic, disjunct and centric distribution patterns

Introduction

Endemics are taxa which occur only in a limited area, for example in the British Isles or in Fennoscandia. **Disjuncts** are taxa distributed in different areas with considerable gaps in between so that genetic exchange is impossible between the separated populations. **Centric distribution** relates to limited areas with an extraordinary concentration of endemics and/or disjuncts.

The problems of interpreting endemic or disjunct distribution patterns are, in many ways, very different from the problems of understanding the distribution patterns of the more widespread taxa dealt with previously. In the latter case it is assumed that the distributional area is not limited by the ability of the taxon to spread and that the distribution is therefore limited by environmental factors. This assumption is supported by fossil evidence. Such an assumption cannot be made with respect to endemics and disjuncts. Instead it seems more probable that an explanation of their distributions is to be sought in the history of the populations, namely that they are **relicts** from earlier, different distribution patterns in the past when the plants could migrate under ecological conditions different from those of today. Such hypotheses should, however, be supported by fossil evidence (Godwin 1975).

In Chapter 4 the history of the flora was outlined based on fossils and geological evidence. The important events to which the flora had to adapt were the Pleistocene glaciations, when ice sheets covered most of Fennoscandia and the British Isles. The glaciations were associated with major climatic depressions and many of the species present in the North European flora, especially the thermophilic and the atlantic species, could not survive closer than in the Mediterranean, Iberia, the Balkans, or south and east of the ice sheets in Russia. Alpine and boreal species that are better adapted to life under colder conditions could survive closer to the ice sheets and, perhaps, on unglaciated areas along the Atlantic coasts of Britain and Fennoscandia.

The hypothesis that parts of our present flora are survivors from pre-glacial or interglacial times in unglaciated enclaves, is called the **nunatak** or the **hibernation hypothesis**. The enclaves may be coastal areas or nunataks. *Nunatak* – which has become an international term – is, in the Eskimo language, a mountain completely surrounded by an ice sheet. The alternative hypothesis, that all biota were exterminated by the glaciations, has been called the **tabula rasa hypothesis**; *tabula rasa* means clear table.

In the following discussion of the endemic, disjunct and centric distribution patterns I shall use the outline of the geological history as given in Chapter 4, namely that there were areas for survival of plants not only south and east of the Pleistocene ice sheets, but also in unglaciated enclaves in the British Isles and along the Atlantic seaboard of Ireland, Scotland and Norway. In or near such refuges it is even possible that some taxa could be descendents of pre-Pleistocene populations.

After presenting the phytogeographic picture that presumes not only the existence of unglaciated refuges but also plant survival in such areas in Northern Europe during the Pleistocene, I shall briefly comment on alternative phytogeographical explanations based on the tabula rasa hypothesis and biological considerations. Also, however, these must be coherent with geology and meteorology, which have recently accumulated much evidence in favour of the existence of unglaciated areas with climatic conditions suitable for the survival of subalpine, alpine and arctic biota (Dahl 1990b, 1991, 1992a).

Ecology of periglacial survival

It has been argued against the nunatak hypothesis that survival of a considerable number of species in limited refuges along the Atlantic Ocean under Ice Age conditions would be impossible for ecological reasons, especially for highly oceanic bryophytes and lichens. It is therefore necessary to consider if the ecological conditions in such refuges could have supported plant life.

Palaeoecological studies in southern England indicate that the climate must have been similar to what is observed today around the timber-line in Northern Europe. The climatic reconstructions combined with fossil evidence suggest that along the southern edge of the ice sheet, where we know that plants survived during the late Weichselian maximum, summers were about 6 °C colder than today, while winters were considerably colder (pp. 32–3). There is no reason to suppose that the difference between present-day conditions and Ice Age conditions was greater along the coasts of northern Europe than south of the ice sheet.

Gjærevoll & Ryvarden (1977) investigated the flora on the Jensen Nunataks in southwest Greenland at an altitude of about 1400 m and found 62 species of vascular plants and numerous mosses and lichens. There were 17 Amphi-Atlantic vascular plants, for example *Carex scirpoidea* (Fig. 81, p. 145) and *Papaver radicatum* coll., and many of the species found have a centric distribution in Scandinavia today, for example *Campanula uniflora*, *Erigeron humilis* and *Carex bicolor*. It is reasonable to assume that present-day conditions on the Jensen Nunataks compare well with the conditions during the coldest episode of the last glacial period on the south Norwegian nunatak Gjevilvasskammen in the Dovre–Trollheimen area, at about 1350 m altitude. The geological reconstruction gives an altitudinal limit of the ice sheet of about 1400 m in the Dovre–Trollheimen area. This area supports a rich alpine flora today with many endemics and species with disjunct or centric distribution patterns (Dahl 1990b).

Most of the discussion concerning survival of plants in unglaciated areas has hitherto been based on the distribution patterns of vascular plants, while bryophytes and lichens have played a subordinate role. The reason is that the distributions of vascular plants have been better known than the distributions of bryophytes and lichens. But during recent decades the study of the geography of bryophytes and lichens has made great advances, and the time has come to begin to use this information. Especially under cold conditions, the number of species of bryophytes and lichens far exceeds the number of vascular plants and this might give a better basis for conclusions. If a vascular plant has survived in an arctic refuge, we can be quite sure that many bryophytes and lichens survived too. Being poikilohydric, the bryophytes and the lichens are especially dependent on humidity factors and this might throw more light on climatic conditions in the refuges.

During the late Weichselian maximum, sea-surface temperatures around the Macaronesian islands were approximately the same as today (CLIMAP Project Members 1976) and near the coast of southwest Spain about 2 °C below present-day values. Air temperatures were lower in Northern Europe both during summer and winter (see pp. 32–3). This would create a moist climate along the coast with high atmospheric humidity, a climate suitable for oceanic poikilohydric plants. Similar situations can be found today along the coast of southernmost Alaska where glaciers come down to the sea and there is a rich flora of oceanic lichens (Krog 1968). That temperate and oceanic bryophytes can survive under arctic conditions is shown by the presence of such species in unglaciated parts on the arctic slope of the Brooks Range in Alaska (Steere 1965) under climatic conditions far more severe than existed along the Atlantic shore of Northern Europe during the Pleistocene. Examples include *Herbertus aduncus*, *Frullania tamarisci*, *Radula complanata* and *Tetraphis pellucida*.

Endemics

The study of endemism has always been central in phytogeography. A high percentage of endemism generally indicates an old flora, and vice versa, a low percentage indicates a young flora. One reason is that the evolution of taxa by genetic drift and selection takes time. If we knew the speed of the evolutionary clock, we could assign an age to the different endemic populations. However, especially concerning evolution of small populations under boreal or arctic conditions, there is no general agreement about evolutionary rates and opinions differ widely. It is possible by a number of techniques to measure genetic distances between populations. If we are able, using other independent evidence, to assign an age to the geographical separation of populations, we might be able to make an estimate of the rate of genetic differentiation.

It is striking that endemic taxa are poorly represented in Northern Europe. Further south, in the mountains of the Alps, Balkans, the Italian and Iberian peninsulas, and on the islands in the Mediterranean there are numerous highly distinctive local endemics, even at a generic level (Farvager 1972; Anonymous 1983). It was long thought that endemic species were missing in the north and this was taken as support for the tabula rasa hypothesis. Later research has shown that this notion at least has to be modified and that subspecific endemism in some phytogeographic elements, notably the subalpine-alpine elements can be quite high.

It is necessary to bear in mind, however, that differentiation of taxa can take place in different ways. One way, termed gradual speciation (Valentine & Løve 1958), is by genetic drift and selection where mutants accumulate in different populations and ultimately give rise to taxonomic differences. Differentiation can also occur, for instance, by polyploidy, by exchange of whole chromosomes or chromosome segments, or, when different populations cross, by combinations of new genomes. This has been termed quantum speciation or abrupt speciation (Valentine & Løve 1958), and can occur suddenly.

A Vascular plant endemics

1 Fennoscandian endemics

Initially I shall consider the pattern of endemism in Fennoscandia, here taken as the area of Norway, Sweden and Finland. We shall also include the Kola Peninsula since there are no barriers for the migration of alpine, subalpine and boreal plant populations between Kola and North Fennoscandia. I shall include all species and subspecies apart from polymorphic, apomictic genera such as *Sorbus, Rubus, Taraxacum* and *Hieracium*. The problems of Fennoscandian endemics have been discussed by Nordhagen (1931, 1935), Knaben (1959, 1979, 1982), Borgen (1987), Dahl (1987, 1989), Nordal (1987) and Jonsell (1990). In addition there are also endemics on Svalbard which have been discussed in relation to the Pleistocene history there by Odasz (1991). We shall list as endemics all species and subspecies and a few taxa of lower rank with a restricted distribution.

The Fennoscandian vascular plant endemics can be divided into four distributional–ecological groups.

1 **Alpine–northern boreal (subalpine) endemics**
2 **Halophilous endemics** (with four subgroups)
3 **Öland–Gotland endemics** (or Baltic island endemics)
4 **Other lowland endemics**

The taxa of the first group grow mainly in the northern boreal and alpine areas, occasionally penetrating to the southern boreal region. Many occur at low elevations along the Atlantic coast and in the north.

The halophilous endemics are restricted to soils influenced by salt from the sea. Included in the halophilous group as a whole are a few species that also occur inland, for example *Leymus arenarius* and *Silene maritima* (see Hultén 1971a). The halophilous endemics can be grouped in the following subgroups:

(a) *Baltic endemics*, mostly brackish water taxa, restricted to the shores of the Baltic including some along the north coast of Germany.
(b) *Baltic–West Scandinavian endemics* occurring both in the Baltic and along the west coast of Fennoscandia.

(c) *North coast endemics* restricted to the northern shores of Fennoscandia.

(d) *South Swedish and Danish sand-dune endemics.*

The members of the third group of endemics are, in Fennoscandia, restricted to the Baltic islands of Öland and Gotland and adjacent islands along the coast of Estonia.

Lists of taxa included in these groups of endemics are given in Appendix IV (part A).

Table 1 shows the total number of species and the total number of endemic species and subspecies in the different elements along with the percentage of endemism. In these enumerations only species considered native are included (see p. 154).

Table 1. *Endemic vascular plant taxa of Fennoscandia*

	Total number of species	Endemic taxa	% endemic
Alpine–northern boreal taxa	244	28	11.5
Halophilous taxa	104	21	20.2
Öland–Gotland taxa	41	11	26.8
Other lowland taxa	1066	5	0.5
Total	1455	65	4.5

It is seen that the subalpine-alpine, the halophilous, and the Öland–Gotland elements are over-represented in endemics while the other elements are under-represented. The figures are statistically highly significant. How can this be explained?

The subalpine-alpine taxa, surviving the environmental conditions in unglaciated refuges, may have had a longer genetic continuity and time for differentiation than the lowland elements immigrating after the Weichselian glaciation.

The halophilous endemics are naturally restricted to seashores and the Baltic is the only large brackish water estuary in North Europe. Wherever brackish water species survived during the Pleistocene glaciations, their only refuge in Northern Europe today is along the shores of the Baltic. Segerstråle (1957) has discussed the immigration of fresh-water and brackish water animals in the Baltic and came to the conclusion that they survived the last glacial age in water bodies east of Finland but west of the Urals, where considerable areas between the White Sea and the Urals were ice-free during the late Weichselian maximum. Climatic reconstructions suggest a relatively moderate depression of summer temperature in this area.

Of special interest among the Baltic endemics is *Euphrasia bottnica* with a very isolated taxonomic position. It belongs to the section Minutiflorae (Pugsley 1936; Yeo 1978), whose other members are to be found in eastern North America. It does not hybridise with other *Euphrasia* species in nature, and it has not been possible to obtain seeds by crossing with other species (Karlsson 1986). *Deschampsia bottnica* belongs to a group of species with other local endemics such as *D. wiebeliana* on the shores of rivers in northwest Germany, *D. littoralis* along the shores of the large Swiss lakes, and *D. obensis* in northern Russia.

The Öland and Gotland endemics are probably relics from the flora which dominated the vegetation in Northern Europe during the cold stages of the Weichselian glaciation, persisting into the late-glacial (Iversen 1954; Godwin 1975).

2 Endemics in the British Isles

The endemics in the British Isles can be subdivided into a similar number of elements (Walters 1978). The first is a **montane and alpine element**. The taxa of this element are restricted to higher altitudes in Wales and England, and often descend to low elevations along the coasts of western Ireland and the west and north coasts of Scotland. This corresponds to the element discussed by Conolly & Dahl (1970). There are **coastal endemics** which are restricted to coastal saline soils or coastal dunes or cliffs. The last element contains the **endemics of the rest of the flora**. The enumeration (Table 2) includes only native species and does not include highly polymorphic, apomictic groups such as *Rubus*, *Sorbus*, *Limonium*, *Taraxacum* and *Hieracium*. A list of the endemic taxa is given in Appendix IV (part B), based mainly on standard floras and atlases. Interestingly 20% of the endemic taxa belong to the genus *Euphrasia*.

Table 2. *Endemics in the British Isles*

	Total number of species	Endemic taxa	% endemic
Montane-alpine taxa	168	32	19.0
Coastal taxa	103	8	7.8
Other taxa	1161	16	1.4

It is seen that the percentage of endemism is high among the montane-alpine taxa, intermediate among the coastal taxa, and low for the rest of the flora.

During the glaciations, plants survived in England south of the glacial limit (see Fig. 8, p. 29). Godwin (1975, p. 431f.) emphasises that many species have been found from earlier interglacials and also continuously during the different Weichselian phases, giving us an effective view of the flora which persisted the last glacial age in England. Besides low-alpine and subalpine species, there are many other cold-tolerant plants in this category, for example *Caltha palustris*, *Ranunculus repens*, *R. lingua*, *R. sceleratus*, *Filipendula ulmaria*, *Menyanthes trifoliata*, *Alisma plantago-aquatica*, *Potamogeton natans* and *Scirpus lacustris*. *Betula pubescens* is nearly continuous during the Weichselian and was probably present in earlier interglacials. Of special interest is the endemic *Linum perenne* ssp. *anglicum* which can be recognised by its seeds and pollen (Godwin 1975, p. 167). It has been identified from earlier interglacials and also from Weichselian and early post-glacial deposits and is almost certainly a glacial survivor in southern England. Thus there were possibilities for the existence of quite a rich flora in England during the Weichselian glaciation and this accounts for a rather high percentage of endemics, compared with Fennoscandia.

The high percentage of endemism among montane and alpine taxa is explained by assuming that these elements survived the last and possibly earlier glacial ages in refuges. Bennett (1984) suggested that *Pinus sylvestris* survived the last glacial age in Ireland and Kinloch *et al.* (1986) concluded, based on fossil evidence and analysis of terpenes, that pine may have survived in northwest Scotland.

3 North Atlantic endemics

There are close relations between the oceanic and the montane-alpine floras in the British Isles and Fennoscandia, and also close ties with the flora of the Faeroes and Iceland. Many species are endemic to this region, especially among the bryophytes and lichens. Of the vascular plants, *Arenaria norvegica* (Fig. 67) and *Papaver radicatum* (with 56 chromosomes) are of particular interest. A list of these endemics is given in Appendix IV (part C).

Fig. 67. Distribution of *Arenaria norvegica* (= *A. ciliata* ssp. *norvegica*) in the Northern Hemisphere. After Hultén & Fries (1986).

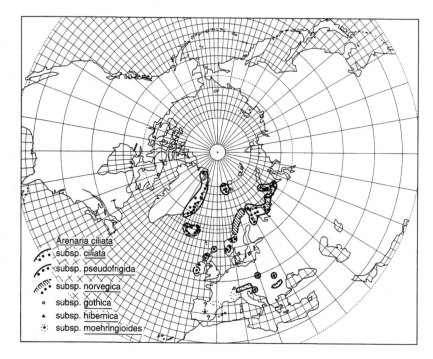

General remarks about European lowland endemism

Dahl (1989) presented a list of European lowland endemics, confined to low elevations in northern France, Belgium, The Netherlands, Germany and Poland. Of these, eight taxa are restricted to heavy metal or serpentine soils where their occurrence tells little about the possibility of their survival during glacial conditions. Three are anthropochorous of unknown origin and nine are dry-land taxa and possibly relicts from the periglacial flora. Of the remaining two taxa, one is an apomict, the other an allopolyploid.

Thus it appears that the lowland flora of Northern Europe that immigrated after the last glacial age is very poor in endemics. The same applies to the prairie states of USA (Gentry 1986) and Canada (personal observations).

The cooling of the climate during late Weichselian time evidently had also an impact on the Mediterranean flora. From the data presented by Gamisans (1976–8), it appears that vegetation types confined to altitudes below 1000 m are very poor in endemics while more cold-tolerant vegetation types have a high percentage of endemics. There is, as far as I know, no endemic on the large islands of the Mediterranean that is exclusive to levels below 1000 m, apart from specialised taxa along the seashores. This probably is an after-effect of the glacial temperature depression.

Dahl (1987) offered the following statement for falsification:

> There is no known instance where it can convincingly be demonstrated that a new amphimictic vascular plant species has evolved by purely genetic processes other than alloploidy during the last 18 000 years.

B Poikilohydric endemics

The poikilohydric plants exhibit distribution patterns different from those of the stenohydric vascular plants. The distribution patterns of oceanic lichens and bryophytes have been discussed (pp. 48–52) especially in relation to humidity factors.

The distributions of bryophytes and lichens are less well known than those of vascular plants, but some species exhibit widely separated disjunctions. Hence, it is possible that some of the endemics listed (Appendix IV) may yet turn up in some remote corner of the world. Several species have been described quite recently and their distribution is insufficiently known. Nevertheless I believe that the information available is sufficient to give a general overall picture.

Important sources of information about the distribution of bryophytes are Ratcliffe (1968), Nyholm (1954–69, 1987–9), Arnell (1956), Smith (1978, 1990), Schuster (1966–80), Daniels & Eddy (1990), Hill *et al.* (1991–4), and for lichens Purvis *et al.* (1992) besides numerous more specific sources. Among the lichens only macrolichens have been considered, since the geography of the crustose lichens is much less well known.

The number of endemics among poikilohydric plants in the North Atlantic region is higher than in vascular plants and often includes easily recognizable taxa. In Appendix IV lists are given of endemic liverworts, mosses and lichens in different parts of the region. A few species with a very restricted distribution also on the French side of the English Channel are included. These endemics fall into different groups.

Some taxa listed as endemics are restricted to lowland Britain (*Cephaloziella nicholsonii, Telaranea murphyae, Ditrichum cornubicum, Weissia mittenii* and *W. multicapsularis*) and may have been introduced by man from unknown countries in a similar way as some mysterious *Bromus* species, or they may have survived the Ice Ages in southern England.

Some endemics are restricted to the Baltic islands (perhaps extending to the Swedish mainland) and may be late-glacial relicts. In this group we find *Plectocolea gothica* and *Tortella rigens*.

A number of the endemics are highly oceanic taxa largely restricted to Ireland, Wales and Scotland. Examples include *Fossombronia fimbriata* (Fig. 68), *Plagiochila atlantica, P. britannica, Fissidens celticus* and

Fig. 68. World distribution of *Fossombronia fimbriata*. After Hill *et al.* (1991).

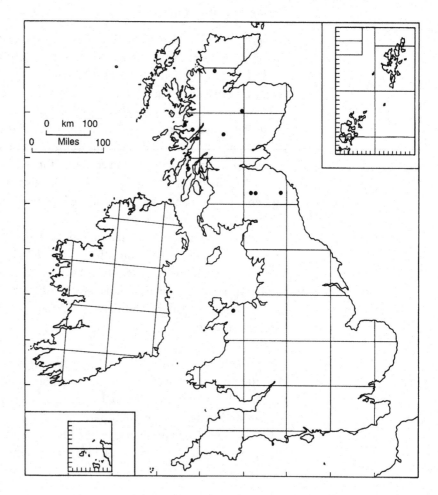

Fig. 69. Distribution of the endemic *Pilophorus strumaticus* (= *P. distans*) in northern Europe. After Ammann & Ammann (1969). Also found locally in northern and western parts of the British Isles (Purvis & Coppins, 1992).

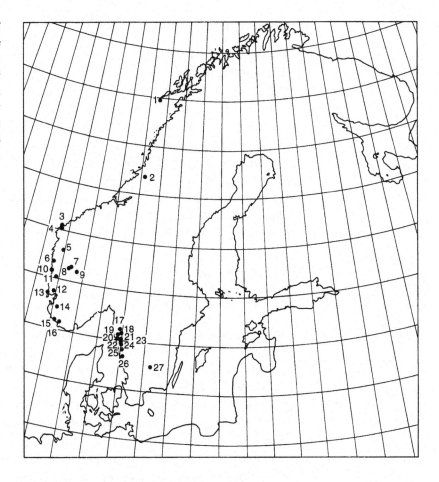

Gymnostomum insigne. Many additional oceanic endemics also grow in western Fennoscandia. Examples among the bryophytes are *Herbertus aduncus* ssp. *hutchinsiae*, *H. borealis*, *Lepidozia pearsonii* and *Weissia perssonii* and, among the lichens, *Pilophorus strumaticus* (Fig. 69) and *Leptogium britannicum*.

Another group of endemic poikilohydric plants includes montane-alpine species. Some of these are restricted to the British Isles, especially Scotland, for example *Bryum dixonii*, others to Fennoscandia, for example *Marsupella andraeoides*, *M. brevissima*, *Scapania sphaerifera*, *Bryum axel-blyttii*, *Tetraplodon blyttii* (also occurring on Svalbard), and *Orthothecium lapponicum*.

Disjunct and relict occurrences

The glaciations changed the ecological conditions and created possibilities for plant migration different from those existing today. It is to be expected that these migrations can be traced in the present-day patterns of plant distributions.

Since the climatic conditions varied in the different glacial and post-glacial periods, disjunct and relict occurrences may have different histories. The origin of disjunct and relict occurrences among alpine – montane and boreal plants probably go back to colder glacial and late-glacial periods, while such occurrences among atlantic and thermophilic plants may have resulted from post-glacial periods with higher summer and winter temperatures.

I European disjuncts

1 Disjunctions of montane-alpine plants

Relatively few of the native, cold-climate vascular plants in Northern Europe are restricted to Europe. Most of them have a circumpolar distribution and several occur in restricted areas on both sides of the Atlantic Ocean (the Amphi-Atlantics).

Some of the montane-alpine taxa restricted to Europe have disjunct distributions, occurring in the north and in the central European mountain ranges.

Alpine plants tend to be limited by high summer temperatures (Chapter 8). During the late Weichselian maximum, summer temperatures south of the glaciated border were about 6 °C lower than today. This means that all plants which today are limited by a maximum summer temperature isotherm of +23 °C or higher had a corridor between Fennoscandia, Britain, and the Alps over the North European lowlands. Subsequently, when the temperatures rose and became too high in the lowlands for many alpine and arctic plants, these became isolated in the central European mountains and the mountains of Fennoscandia and Britain. Fossils of alpine plants in lowland deposits of Weichselian age provide evidence that such migrations actually took place (H. H. Birks 1994). This migration route would, however, not be open to species of alpine plants not tolerating summer temperatures above +23 °C.

There are more European montane-alpine plants in common between Fennoscandia and Britain than with either region and the Alps. The following 24 are common to Britain and Fennoscandia and do not occur in the Alps:

Luzula arcuata	*Minuartia rubella*
Juncus biglumis	*Cerastium arcticum*
Poa flexuosa	*Sagina intermedia*
Festuca vivipara	*Draba norvegica*
Deschampsia alpina	*Saxifraga stellaris* ssp. *stellaris*
Carex rariflora	*S. rivularis*
C. saxatilis	*Phyllodoce coerulea*
Salix myrsinites	*Euphrasia arctica*
S. lanata	*E. frigida*
S. arbuscula	*E. scottica*
Koenigia islandica	*Rhinanthus groenlandicus*
Arenaria norvegica	*Erigeron borealis*

British montane-alpine species occurring in the Alps, but missing in Fennoscandia, amount to the following nine:

Cochlearia pyrenaica
Minuartia sedoides
Oxytropis halleri
Saxifraga rosacea
Meum athamanticum
Myosotis alpestris
Lloydia serotina
Sesleria albicans
Gentiana verna

Similarly, there are only six Fennoscandian alpine species also occurring in Central Europe that are missing in Britain:

Ranunculus platanifolius
Gentiana purpurea
Myostis decumbens
Campanula barbata
Chamaeorchis alpina
Nigritella nigra

The comparisons emphasise the close floristic relationships across the North Sea.

2 Disjunctions of boreal plants

Boreal plants tend to require low winter temperatures (Chapter 7). According to climate reconstructions, winter temperatures were strongly depressed in Northern Europe during the late Weichselian maximum, by at least 10–15 °C. This created favourable conditions for the migration of boreal plants. Fossils of boreal species like *Betula nana, Polygonum viviparum* and *Ranunculus hyperboreus* in Europe outside their present range show that boreal plants had possibilities for migration during glacial and late-glacial times. Thus a whole group of boreal species occur as disjuncts in the Massif Central in France and re-appear in the Alps:

Betula nana
Salix lapponum
Saxifraga nivalis
Alchemilla glomerulans
Vicia dumetorum
Viola mirabilis
Trientalis europaea
Omalotheca norvegica
Cicerbita alpina
Carex chordorrhiza
C. pauciflora
C. brunnescens
C. vaginata

Several boreal species have disjunct occurrences within the British Isles. Examples are:

Woodsia alpina
Cystopteris montana
Rubus chamaemorus
Vaccinium microcarpum
Cornus suecica
Polemonium coeruleum
Pinguicula alpina (extinct)
Linnaea borealis
Cicerbita alpina
Polygonatum verticillatum
Carex ornithopoda
Alopecurus alpinus

They tend to have an eastern and continental distribution pattern in Britain. A quite peculiar distribution is shown by *Woodsia alpina* in Wales. Elsewhere the species is limited by the −6°C isotherm for the coldest winter months, while the highest summits in Wales have winter temperatures around only 0°C.

3 Disjunctions of atlantic plants

Atlantic plants are sensitive to low winter temperatures (Chapter 5). During parts of the post-glacial period winter temperatures may have been approximately 4°C higher than presently measured by the temperatures of the coldest month in the lowest parts of the landscape. This enabled atlantic plants to migrate, for example, between southern Scandinavia and western Norway or between France and Ireland. After the subsequent climatic deterioration populations in between died out, thereby forming a disjunct pattern. Examples of such atlantic disjunctions in western Norway are:

Asplenium adiantum-nigrum
A. marinum
Scilla verna
Saxifraga hypnoides
Alchemilla xanthochlora
Vicia orobus
Polygala serpyllifolia
Callitriche pedunculata
Conopodium majus
Erica cinerea
Lysimachia nemorum
Euphrasia scottica
Centaurea pseudophrygia
Scirpus mamillatus ssp. *austriacus*

Some of these, for example *Euphrasia scottica*, may have immigrated across a land connection in the North Sea in early post-glacial times as has also been suggested for *Ulmus* in Norway by Lindquist (1932). Examples of such disjuncts in Ireland and southwest England are:

Arbutus unedo
Daboecia cantabrica
Erica ciliaris

E. erigena
E. vagans
Neotinea intacta

4 Late-glacial relicts

Fossil evidence has given a picture of the flora of the periglacial areas during the Weichselian maximum and during the late-glacial (Iversen 1954; Godwin 1975). Remnants of this flora are today found in a disjunct pattern including western Ireland, the Pennines in England, and the Baltic islands. These areas have a high number of endemics, and in addition many disjuncts whose restricted occurrences are relicts of a wider, more continuous distribution. Examples of late-glacial relicts in the British Isles include:

Viola rupestris
Polygala amara
Tuberaria guttata
Helianthemum canum
Potentilla fruticosa
Primula farinosa
Gentiana verna
Euphrasia salisburgensis

Examples from the Baltic islands with outliers in south Sweden are:

Arenaria gothica
Potentilla fruticosa
Geranium lanuginosum
Helianthemum oelandicum
H. canum
Bartsia alpina
Euphrasia salisburgensis var. *schoenicola*
Pinguicula alpina
Plantago tenuifolia
Artemisia laciniata
Saussurea alpina ssp. *esthonica*

In addition several bryophytes and lichens with disjunct occurrences in Öland and Gotland, for example *Pseudocalliergon trifarium*, *P. turgescens*, *Physconia muscigena* and *Thamnolia vermicularis*, and the endemic taxa discussed on p. 120 are interpreted as late-glacial relicts.

Many of the species listed above are typical of dry calcareous communities described by Krahulec *et al.* (1986). According to Berglund (1966a), *Bartsia alpina* and *Saussurea alpina* were already present in late-glacial times in southern Sweden. According to Molau (1991), *Bartsia* in Gotland belongs to a Central European biotype differing from the biotypes in the mountains of Fennoscandia. There are also outlying populations of *Lychnis alpina* in Öland and southeast Norway which Bøcher (1963) described as a subspecies different from the normal alpine population. Genetically they differ from the normal populations of the species in the mountains but they compare well with serpentine races (Haraldsen & Wesenberg 1993).

Besides dry-habitat species, several taxa are typical of calcareous mires. In the periglacial areas with loess formation the mires were almost exclusively calcareous as shown by fossil mosses from such deposits (Rybníček 1973). Acid indicators like *Sphagnum* species are rare or missing.

Another type of disjunction, probably of late-glacial origin, is shown by vascular plants such as *Diplazium sibiricum*, *Cystopteris sudetica*, *Clematis sibirica* and *Aster subintegerrimus* in the upper valleys of southern Norway. Their main distribution areas are eastern, with western limits in eastern Finland and the Baltic. They all have light, wind-dispersed diaspores, giving possibilities for rapid colonisation after the retreating ice.

Another group detected by Ahlner (1948) consists of lichens restricted to the very dry soils in Gudbrandsdalen in central South Norway. Some have their nearest stations in the Baltic islands, for example *Psora vallesiaca*, *Heppia lutosa* and *Squamarina lentigera*, some in Central Europe, for example *Solorinella astericus*, *Toninia tristis* and *Phaeorrhiza sareptana*, and some in southeast Europe, for example *Caloplaca tominii*. Some of these have disjunctions also to the Arctic, for example *Psora vallesiaca* and *Squamarina lentigera*.

5 Disjunctions of thermophilic plants

As shown by fossil evidence (p. 80) several thermophilic plants occurred further north than today and timber-lines were raised, indicating higher summer temperatures some time during the post-glacial; the difference amounts to about 2 °C or about 0.2 units on the R-scale (see Fig. 29, p. 68).

This enabled thermophilic plants to migrate northwards. When summer temperatures during the later Holocene fell, some persisted in favourable areas north of their main distribution.

Especially on the Baltic islands a number of disjunctions of more thermophilic species occur (that hardly can be interpreted as late-glacial relicts discussed above). Examples are:

Ceterach officinarum
Calamagrostis varia
Stipa pennata var. *joannis*
Orchis spitzelii
Ranunculus illyricus
R. ophioglossifolius
Coronilla emerus
Viola alba
Fumana procumbens
Globularia vulgaris
Aster linosyris
Lactuca quercina
Tragopogon crocifolius

Some disjunct thermophiles also occur along the south coast of Norway, for instance *Coronilla emerus* and *Silene armeria* (Bronger 1992). In addition *Tilia platyphyllos* has outliers along the west coast of Sweden, and *Ligustrum vulgare* in an area around the Oslofjord in South Norway. *Ligustrum* was already present in south-central Sweden during the Younger Dryas period (Florin 1979), but has since become extinct. *Mercurialis perennis* has a peculiar disjunction from South Norway to Nordland (North Norway).

II Extra-European disjunctions

The stenohydric and the poikilohydric plants exhibit rather different distribution patterns. Among the stenohydric, mainly vascular plants, the thermophilic and atlantic species are generally endemic to Europe and adjacent parts of Africa and Asia. The more cold-tolerant species often have a circumpolar or Amphi-Atlantic distribution. Very few have a bipolar type of distribution. Among the poikilohydric bryophytes and lichens the oceanic species very often have extra-European disjunctions to Macaronesia, to southeast North America, to the Tropics, to southeast

Asia, and to the Pacific. Also an Amphi-Atlantic distribution tends to be the rule rather than the exception. Of the more cold-tolerant species quite a number have a bipolar distribution. In general, poikilohydric species have wider distributions than vascular plant species, reminiscent more of the pattern exhibited by genera and families of vascular plants. This may, in part, have ecological causes, but it is also possible that the distribution of lichens and bryophytes must be understood on the basis of their considerably older geological history than the vascular plant species. Some authors have suggested that such disjunctions date back to Mesozoic times.

1 Amphi-Atlantic disjuncts

One feature which already attracted the attention of Blytt (1876) was the presence in Norway of a number of species found west of the North Atlantic Ocean but restricted to Fennoscandia on the European side. This has been termed a **West Arctic element** in the flora of Fennoscandia (cf. Table 3). Blytt observed that such species tended to occur in more continental parts of the mountain chain, east of the highest mountains, and to avoid mountains with a more oceanic climate. To explain this pattern Sernander (1896), as the first, proposed that a 'not inconsiderable' number of Fennoscandian alpine plants had survived the last glacial age in refuges along the Atlantic coast of Norway, and that they were relicts from an inter-glacial or a pre-glacial distribution pattern. A similar hypothesis had previously been proposed for the Greenland flora by Warming (1888). The hypothesis, known as the nunatak hypothesis, has been at the centre of a heated debate for more than 100 years.

The floristic connection across the northern Atlantic Ocean has attracted the attention of numerous authors. The problem was discussed and reviewed during a symposium in Iceland (Løve & Løve 1963) and since then the development of the discussion has been reviewed by Dahl (1987, 1991).

Iceland has very clearly a European flora despite its isolation by the sea from Europe. There are a few American taxa on Iceland, with an over-representation of species adapted to long-distance dispersal. This applies also to insects. The southeastern coast of Greenland has strong European connections, and it has been discussed whether the phytogeographic limit between Europe and America should be drawn along the Greenland inland ice, or perhaps along the Davis Strait (Dahl 1963b).

Table 3. *West Arctic vascular plant species*

Salix glauca ssp. *callicarpaea*	*Pedicularis flammea* (Fig. 70)
Sagina caespitosa	Erigeron humilis
Minuartia rossii (Fig. 73)	Antennaria porsildii
Arenaria humifusa	Carex nardina
Papaver dahlianum	C. scirpoidea (Fig. 81, p. 145)
Draba crassifolia	C. arctogena
D. oxycarpa	C. rufina
Braya linearis	C. macloviana
Potentilla chamissonis	C. stylosa
P. rubricaulis	Phippsia algida ssp. *algidiformis*
Epilobium lactiflorum	Puccinellia phryganodes ssp. *neoarctica*
Rhododendron lapponicum	Leucorchis straminea

The Amphi-Atlantic species are represented on both sides of the North Atlantic. They have a western distribution in Europe and their limits eastwards is to the eastern slopes of the Ural Mountains, and with large gaps further eastwards, perhaps again occurring along the Pacific coasts. In North America the Amphi-Atlantics have an eastern distribution going westwards no further than the Canadian Arctic Archipelago and perhaps re-appearing along the Pacific coast. The distributions of the Amphi-Atlantic vascular plants have been beautifully mapped by Hultén (1958) and by Hultén & Fries (1986).

Among the Amphi-Atlantics there are several different distribution types. Some species have a very restricted area in Europe but are widespread in Greenland and America (cf. Table 3). An example is *Pedicularis flammea* (Fig. 70). Other species have a wide area in Europe and only reach Greenland on the American side (Table 4). An example is *Alchemilla alpina* (Fig. 71). Still others have a more symmetric and wider range, for example *Euphrasia frigida* (Fig. 72). Some have a high-arctic distribution, connecting Svalbard with northern Greenland. An example is *Minuartia rossii* (Fig. 73). Others have a more low-arctic distribution, as shown by *Euphrasia frigida*. Finally, there is an American element in the British Isles of four species (discussed on p. 136). As far as is known no species restricted to the temperate regions of Europe and America shows an Amphi-Atlantic distribution pattern; however, there are several vicariant species on both sides of the Atlantic Ocean.

Based on the above sources and later information I have counted the number of Amphi-Atlantic vascular plant taxa (species and subspecies)

Fig. 70. Distribution of *Pedicularis flammea*, a West Arctic species in Fennoscandia, in the Northern Hemisphere. After Hultén & Fries (1986).

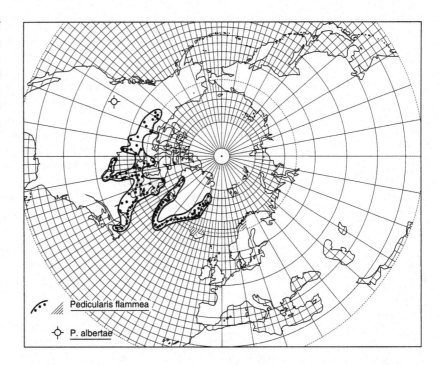

Fig. 71. Distribution of *Alchemilla alpina*, a European species in Greenland, in the Northern Hemisphere. After Hultén & Fries (1986).

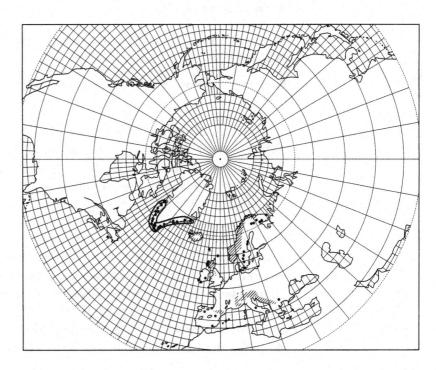

Table 4. *The European element of vascular plants in Greenland*

Isoetes lacustris	*Viola montana*
Botrychium boreale	*Angelica archangelica* ssp. *archangelica*
Athyrium distentifolium ssp. *distentifolium*	*Armeria maritima* ssp. *maritima*
Polypodium vulgare	*Vaccinium myrtillus*
Betula nana	*Gentiana aurea*
B. pubescens ssp. *tortuosa*	*Callitriche hamulata*
Atriplex longipes ssp. *praecox*	*Thymus praecox* ssp. *arcticus*
Arenaria ciliata ssp. *pseudofrigida*	*Veronica fruticans*
Cerastium fontanum ssp. *scandicum*	*Erigeron borealis* s.l.
Ranunculus glacialis	*Antennaria porsildii*
Draba sibirica?	*Cirsium helenoides*
D. oxycarpa	*Hieracium alpinum* complex
Rorippa islandica	*Catabrosa aquatica* ssp. *aquatica*
Braya linearis	*Poa flexuosa*
Sedum acre	*Phippsia algida* ssp. *algidiformis*
S. annuum	*Anthoxanthum odoratum* ssp. *alpinum*
Saxifraga rosacea	*Nardus stricta*
S. paniculata ssp. *paniculata*	*Elymus arenarius* ssp. *arenarius*
Dryas octopetala	*Eleocharis quinqueflora*
Rubus saxatilis	*Carex parallela*
Alchemilla alpina (Fig. 71)	*C. norvegica* ssp. *norvegica*
A. wichurae	*C. panicea*
Geranium sylvaticum	*Juncus squarrosus*
Polygala serpyllifolia	

Based on Bøcher *et al.* 1968, 1978; Feilberg 1984; Hultén & Fries 1986; Bay 1992.

present in the flora of the different parts of the North Atlantic region (Fig. 74). It is seen that the highest number of Amphi-Atlantics on the European side of the ocean is found in northern Sweden and Norway. From there the number drops steadily towards the east and the south. A fair number of Amphi-Atlantics is found in Scotland. On the American side of the ocean the highest number is found in southwest Greenland and the number decreases towards the west and the south. Only a few Amphi-Atlantics are found in the southern Appalachian mountains. Species believed to have been human introductions have been excluded from these enumerations including several species once considered native in Newfoundland but probably introduced by British and Spanish fishermen. They carried ballast in their empty boats on their way across the ocean, deposited the ballast on land, and returned with fish. This has been

Fig. 72. Distribution of
Euphrasia frigida, a
symmetric Amphi-
Atlantic species, in the
Northern Hemisphere.
After Hultén & Fries
(1986).

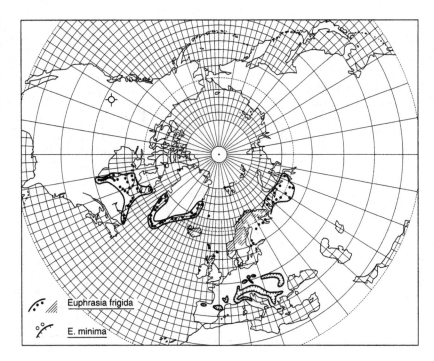

Fig. 73. Distribution of
Minuartia rosii, a high-
arctic West Arctic species,
in the Northern
Hemisphere. After Hultén
& Fries (1986).

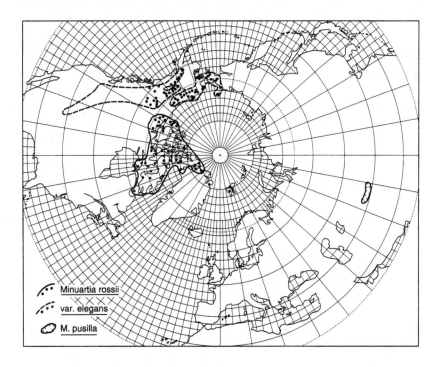

Fig. 74. The number of Amphi-Atlantic taxa in different parts of Europe and North America. Further explanation is given in the text. After Dahl (1963b).

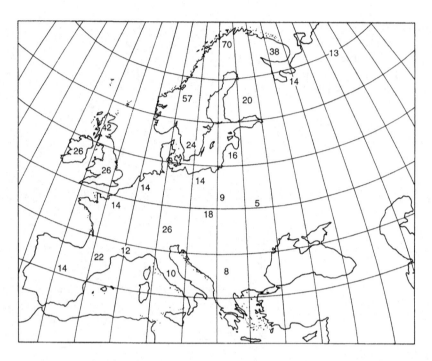

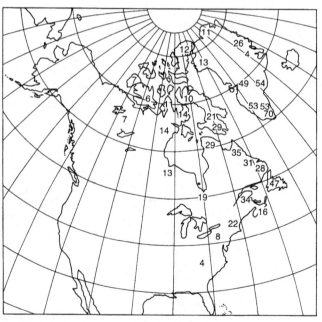

elegantly analysed, for both vascular plants and insects, by Lindroth (1957).

Species belonging to the West Arctic element have their eastern limit in Fennoscandia and Svalbard. They are generally calcicolous and continental, some of them are scree plants such as *Papaver* spp., *Braya linearis*, and perhaps *Arenaria humifusa*. Of the species reaching their easternmost outposts in Svalbard, there are two which might easily have been transported across the sea from Greenland to Svalbard: *Salix glauca* ssp. *callicarpaea*, which consists of a single male individual on Svalbard, and *Puccinellia phryganodes* ssp. *neoarctica*, which is a halophilous species.

On the American side many species reach their westernmost outposts in Greenland, where they form a European element. This element is listed in Table 4.

Bøcher (1951) pointed out that the European element in Greenland consists of many species tending to grow in oceanic areas, often oligotrophic and appearing late in primary successions. He suggested that the European taxa migrating to southern Greenland came by a western and southern route across the North Atlantic. The atlantic climate on this route favoured migration of oceanic and oligotrophic taxa or biotypes (for example *Alchemilla alpina*, Fig. 71). On the other hand, the American species that migrated to Europe, came by a more northern and eastern route over Svalbard to northern Fennoscandia, and here climatic conditions favoured continental and calcicolous biotypes (an example is *Pedicularis flammea*, Fig. 70).

It has been emphasised that the Amphi-Atlantic vascular plants are montane-alpine-arctic. Very few, if any, temperate species exhibit an Amphi-Atlantic pattern. Thus the climate on a land bridge across the North Atlantic must have been cold and much like the present (Dahl 1987). However, such conditions only occurred with the great glaciations during the last few million years. Therefore, the migrations postulated above must be of a relatively recent occurrence. That many Amphi-Atlantic taxa belong to polymorphic groups and may be of low taxonomic rank (Dahl 1987) also suggests a relatively recent separation.

Amphi-Atlantic bryophytes and lichens

Lynge (1938) observed that the macrolichen floras of Svalbard and northeast Greenland were almost identical. Dahl (1950) found a closer correspondence between the macrolichen floras of southwest Greenland and alpine areas of Fennoscandia, than between Fennoscandia and the Alps. Later studies have emphasised this. The bryophyte and lichen floras of Newfoundland and western North Europe are very similar with many corresponding species, especially acidophilous and oligotrophic species such as *Sphagnum angermanicum* (Fig. 75), *Cetraria islandica* ssp. *crispiformis* and *Stereocaulon dactylophyllum*. Among the oceanic bryophytes in Europe, many are also found in the southern Appalachians. Apparently the region on both sides of the North Atlantic forms a floristic region by itself with another Amphi-Beringian region around the North Pacific.

2 The American and Lusitanian elements in the British Isles

The American element in the British Isles has often been discussed (Perring 1967). But the number of taxa has been reduced due to new information and I would only include the following four:

Potamogeton epihydrus
Eriocaulon aquaticum
Sisyrinchium bermudianum
Spiranthes romanzoffiana

Fig. 75. World distribution of *Sphagnum angermanicum*, an Amphi-Atlantic moss species. Contributed by K. I. Flatberg, 1994.

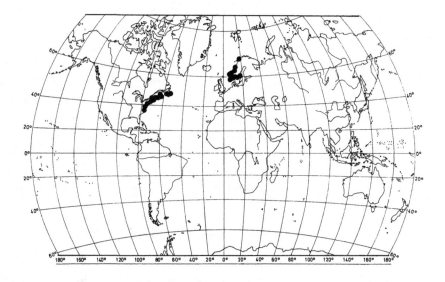

136

Myriophyllum exalbescens, previously included in this element, has been found to be identical to *M. sibiricum* which is almost circumpolar. *Najas flexilis* had a much wider distribution in early post-glacial times than today and occurs at present in scattered localities in northern Europe. *Limosella subulata* grows on mud-flats in Wales and I suspect it may have been introduced by fishing boats from Nova Scotia during early fishing in eastern North America as described by Lindroth (1957). These are all wet mire, lake, or other habitat plants, as are the four species listed above.

Pollen of *Eriocaulon aquaticum* has been found in interglacial deposits in Ireland back to the Gortian interglacial (Jessen *et al.* 1959; Godwin 1975). Apparently the pollen corresponds in size with present-day European populations of the species, while the American population is reported to have smaller pollen. If this is confirmed it suggests that *Eriocaulon*, and probably also the other members of the American element, may have had a very long history in Ireland.

The following Lusitanian species have a disjunct occurrence in the British Isles, re-appearing in the Iberian Peninsula and/or in the Alps:

Minuartia recurva
Saxifraga spathularis
S. hirsuta
Erica mackaiana
Pinguicula grandiflora

In Britain their distribution corresponds to that of the Hibernian group that is suggested to be limited by winter frost (pp. 41–2). But the Lusitanian species tend to ascend to considerable altitudes in the mountains where winters become colder. The element has been discussed by Mitchell & Watts (1970). Fossils of *Erica mackaiana* have been found back to the penultimate interglacial in Ireland and the species may have survived in unglaciated refuges.

3 Macaronesian disjuncts

The Macaronesian disjuncts occur along the western fringes of Europe and then again on the Macaronesian islands – the Azores, Madeira, the Canary Islands and the Cape Verde Islands. I do not include in this element species with any extensive distribution in the Mediterranean region. The best known vascular plant species with this type of distribution is *Hymenophyllum tunbrigense* which is a poikilohydric filmy fern. To the element may be included also *Dryopteris aemula* and *Trichomanes speciosum*. This element includes many oceanic bryophytes, especially hepatics, and lichens. A list of members of the element is given in Appendix V. Many of the Macaronesian bryophytes and lichens have a wide distribution in tropical countries. There are particularly close connections with South America and some occur in the southeastern states of the USA.

4 Southeast Asiatic disjuncts

A number of European bryophytes, especially hepatics, and lichens have disjunctions in the eastern Himalaya, and provinces of China such as Yunnan. A list of such species is given in Appendix V. Some of them also occur in the Pacific area.

The southeast Asian disjuncts are highly oceanic species in Europe. One possible explanation of their distribution pattern would be migrations during the Pleistocene glaciations. But both climate reconstructions and fossil evidence suggest that conditions in temperate and Mediterranean regions were drier during the glacial ages than now. Müller (1954–7) suggested that the last opportunity for migration of oceanic hepatics between western Europe and southeast Asia was along the Tethys Sea, and thus refers back to the Mesozoic. This of course raises the question whether the species could have remained constant for so long and he writes:

> Among the Eocene hepatics, about 60 million years old preserved in amber, are species which compare so accurately with present-day species of *Mastigiolejeunea* or *Ptychocolea* as it is at all possible by a comparison of fossil remains with plants living today. (Müller, reprint edn. 1954–7, p. 237, translated from German)

5 Pacific American–European disjuncts

These are disjunctions between Northern Europe and the western coast of America, from California to Alaska and adjacent mountains. Among the vascular plants the element is small, oddly enough consisting mainly of pteridophytes. The following species have this type of distribution:

Botrychium boreale
Thelypteris limbosperma
Asplenium septentrionale
A. adiantum-nigrum
Polystichum braunii
Blechnum spicant (Fig. 76)
Lemna gibba
Saxifraga adscendens

Among bryophytes and lichens the element comprises many more species. A list is given in Appendix V.

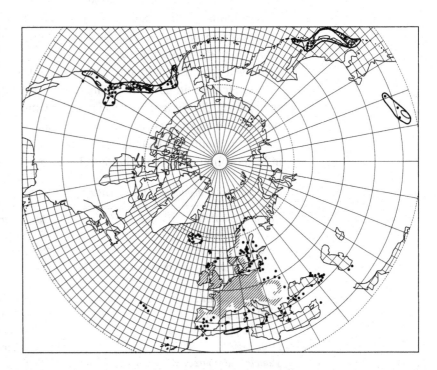

Fig. 76. Distribution of *Blechnum spicant*, a Pacific disjunct, in the Northern Hemisphere. After Hultén & Fries (1986).

6 Bipolar distributions

The problem of bipolar disjunctions has long been discussed in relation to vascular plants, for instance by DuRietz (1940) and Raven (1963). The species can have enormous gaps between an area in New Zealand or in southernmost South America and their nearest occurrences in the Rocky Mountains or the Alps or Himalaya. Many of the bipolar disjuncts have a montane-alpine or arctic distribution in the Northern Hemisphere (Table 5). This is a quite unbalanced element with a high over-representation of graminoids, grasses and sedges.

There have been speculations whether the bipolar distributions could be explained as a result of long-distance dispersal by migrating birds across the Equator in America, perhaps during the Pleistocene glaciations. The distance from the highest mountains in Costa Rica, which carried glaciers during the Pleistocene, to the high mountains in Colombia is not so great, with further access southwards in the Andes. But this would leave us with the question of how to explain the quite numerous bipolar species common to South America and New Zealand. A migration across the Antarctic continent before it broke up could be a possible explanation, corresponding to the standard explanation of the distribution of the Proteaceae, *Nothofagus*, and other austral genera of vascular plants. But such a bipolar migration must have taken place long before the late Tertiary, a conclusion already reached by DuRietz (1940, p. 272).

To stretch the limits of credibility further, several exceedingly local and ecologically discriminating species are involved: *Tetradontium*, *Oedopodium* and *Buxbaumia* are obvious examples.

Still the Southern Hemisphere populations appear to have diverged little morphologically or ecologically from the Northern Hemisphere

Table 5. *Bipolar disjuncts in the vascular plant flora of Northern Europe*

Koenigia islandica	*Phleum alpinum*
Montia fontana	*Trisetum spicatum*
Cerastium arvense ?	*Carex arctogena*
Empetrum nigrum and related taxa (Fig. 77)	*C. capitata*
Plantago maritima	*C. curta*
Triglochin palustre	*C. lachenalii*
Catabrosa aquatica	*C. macloviana*
Deschampsia caespitosa	*C. magellanica*
D. flexuosa	*C. maritima*
Vahlodea atropurpurea	*C. microglochin*

Fig. 77. Distribution of
Empetraceae, a family
with a bipolar disjunction.
After Gjærevoll (1973).

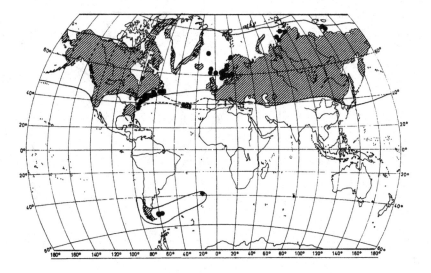

Fig. 77. Distribution of
Empetraceae, a family
with a bipolar disjunction.
After Gjærevoll (1973).

populations. Such an observation may support the idea of a relatively recent origin of this element in the Southern Hemisphere. On the other hand, it may reflect the possibility that these species are simply highly conservative throughout their range. Since asexual reproduction predominates in the Southern Hemisphere populations, this would fix this conservatism.

Less attention has been given to bipolar poikilohydric plants. These are numerous. Schofield (1974) gives a list of 83 bipolar mosses in New Zealand of which 14 are considered introduced. In addition there are 24 native disjunct mosses in South America not present in New Zealand. According to Smith (1990), only 12 British hepatics are bipolar. Schofield (1970, p. 193) lists nine bipolar hepatics in western North America. Based on Galloway (1985) and Purvis *et al.* (1992) as well as other sources, about 60 species of macrolichens exhibit a bipolar distribution. The total element according to these enumerations amounts to about 170 species. Good examples are *Oedopodium griffithianum* (Fig. 78) and *Cetraria delisei* (Fig. 79).

Of the many native bipolar mosses in New Zealand and South America, 57% are monoecious, the rest are strictly dioecious, and 17% are exclusively sterile. Of the rest many are only rarely fertile. Most of the lichens reproduce by thallus fragments, isidia, or soredia. Many are only locally distributed and ecologically specialised. Some of the bipolar species belong to species complexes with many species in the south but with one or a few species in the north, for instance *Neuropogon*

Fig. 78. World distri-
bution of *Oedopodium
griffithianum*, a bipolar
moss. After Schofield
(1974).

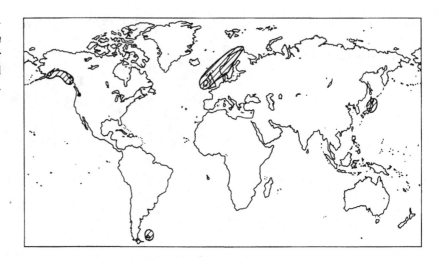

sulphureus, Placopsis gelida and *Siphula ceratites*. Others belong to northern species complexes with a few species in the south, for instance many *Sphagnum, Cetraria* and *Cladonia* species.

If there are difficulties in explaining the bipolar distribution of vascular plants by long-distance dispersal across the Equator, this explanation becomes far more difficult when the more numerous bipolar bryophytes and lichens are considered. Many of them, for instance *Neuropogon sulphureus*, grow today in localities where hardly any far-travelling birds occur. The birds that fly across the Equator from North America to the Andes and southern South America are waders, some web-footed birds, some birds of prey, and some swallows. Many of the lichens grow on stones, for instance *Umbilicaria* species and *Neuropogon sulphureus*, where such birds rarely rest. Birds generally like to rest on top of big stones, where their droppings alter the conditions for lichen growth. This results in the invasion of coprophilous species, but as far as I can see no coprophilous macrolichen is among the bipolar species.

Schofield (1974) makes the following comments on the explanation of the bipolar disjunctions of mosses growing in New Zealand:

Long-distance dispersal seems the only reasonable answer, and the predominance of monoecious condition or vegetatively reproducing successful species tends to support this. It is possible that the origin of the disjunct element is relatively recent, possibly from the Pleistocene or the following glaciations. Since many of these disjunctions are

Fig. 79. World distribution of *Cetraria delisei*, a bipolar lichen. After Kärnefelt (1979).

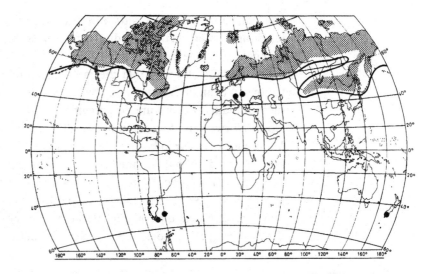

presently unknown from South America it is necessary to assume a remarkably long distance of dispersal from the Northern Hemisphere populations.

Schuster's (1969) analysis of the bipolar hepatics led him to a similar conclusion, although the number of taxa is much smaller. Smith (1972), however, from a detailed study of the Polytrichaceae, concluded that this family originated in Gondwanaland and moved northwards with the drifting of the continental plates. This would make some of the taxa as old as mid-Mesozoic. Scott (1988), discussing Australasian bryogeography, concluded that for widespread taxa, among them the bipolar, patterns of distribution would 'have been influenced largely by a combination of continental drift and drying up of formerly wetter areas, with radiating dispersal over short distances and, less commonly, by long range dispersal'.

In conclusion no reasonable explanation is in sight, and it must be hoped that geology one day may come up with an answer.

Centric distributions

Refugia are areas where plants and animals could survive while they became extinct elsewhere. In or near refugia one expects to find concentrations of disjunct species. One also expects to find a higher degree of endemism than in other areas, since the plants or animals have a longer history in or near the refugia than elsewhere. Inversely, when a remarkable

143

concentration of disjunct and/or locally endemic species is observed it is tempting to interpret this as indicating the presence of refugia during some past geological period (cf. Birks 1993).

Fries (1913) observed that the rare and disjunct plants of the alpine regions in Scandinavia tended to be concentrated in two areas. One is in the mountains of Dovre and Jotunheimen in South Norway, the other is in North Norway and Sweden northwards from the Arctic Circle (Fig. 80). Many species occur in both these areas; these he called **bicentric**. Others are restricted to one of the areas; these were called **unicentric** by Arwidsson (1943). The bicentric species are listed and discussed by Berg (1963). Fries (1913) suggested that the unicentric and bicentric patterns found in Fennoscandia could be explained by survival of plants in or near the two areas, on the coast of South Norway south of the Trondheim Fjord

Fig. 80. The two areas in Fennoscandia with a particularly rich flora of endemic and disjunct mountain plants. After Gjærevoll (1973).

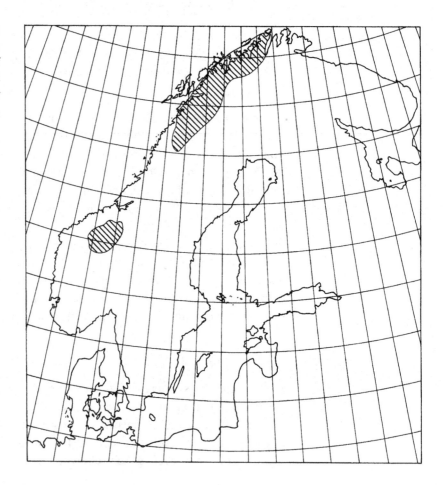

and in the Lofoten and Vesterålen islands in North Norway. In these areas an alpine topography also suggests the presence of nunataks. Centric distribution patterns are also observed elsewhere, for example in Greenland (Gelting 1934, Bøcher 1938, 1951, 1975) and Iceland (Steindorsson 1963). The pattern of disjunct late-glacial relicts discussed above (pp. 126–7) is another example.

If a plant population survives in a limited refuge the result is inbreeding and loss of alleles, especially if the population is kept at low numbers for many generations. It has been emphasised by, among others, Braun-Blanquet (1923), Fernald (1925), and Hultén (1937), that such relict plant populations have little ability to adapt to new conditions. They tend to become ecologically specialised, show little morphological variation, and be unable to invade new territories that become available. This was called **rigidity** by Hultén (1937 p. 20f.). Fernald (1925) pointed out that the disjunct Cordilleran species in Gaspé and Newfoundland were more rigid than their relatives in the western areas.

This phenomenon is clearly represented among the Scandinavian centric plants (Dahl 1991). An example is *Carex scirpoidea* which in Europe is restricted to two mountains not far apart in Norway (Fig. 81).

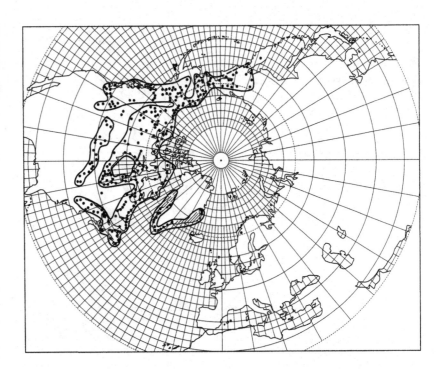

Fig. 81. Distribution of *Carex scirpoidea*, a West Arctic species in the Northern Hemisphere. After Hultén & Fries (1986).

It grows in sites rich in calcium, in snow-free areas in the low- to mid-alpine region. In North America it has a wide distribution occurring in a variety of habitats and is in no way restricted to calcareous parent rocks. Another example is *Rhododendron lapponicum* (Fig. 82), which is bicentric with a somewhat continental distribution and calcicolous. In North America it is found over wide areas growing on granite in New England and even in the Driftless area of Wisconsin with a temperate deciduous vegetation. A third example is *Kobresia simpliciuscula*, which in Greenland is restricted to continental regions with yearly precipitation between 300 and 600 mm and an annual temperature amplitude of at least 20 °C (Bøcher 1938, 1951). In Norway it is strictly northern boreal-alpine while in Teesdale in northern England it grows at altitudes where oak also can grow and with much milder winters than in Greenland. The late Professor Tyge Bøcher told me that he could grow the Teesdale population outdoors in Copenhagen while the Greenland populations could only be maintained in the arctic greenhouse.

Perhaps it is a case of rigidity that the British endemic *Primula scotica* on mainland Scotland and the Orkneys is restricted to areas that I consider unglaciated during the Weichselian glaciation.

Fig. 82. Distribution of *Rhododendron lapponicum*, a bicentric, West Arctic species in Fennoscandia, in the Northern Hemisphere. After Hultén & Fries (1986).

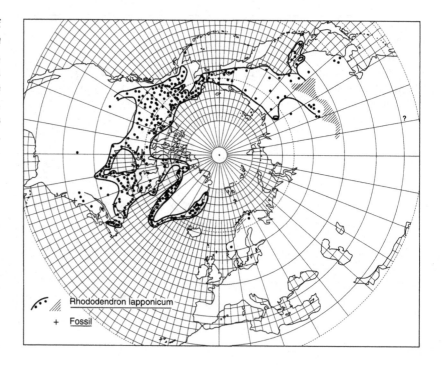

Alternative phytogeographic explanations

The question of plant and animal survival on ice-free refugia is by no means unique to Northern Europe, but pertinent to all the coasts bordering the North Atlantic Ocean (cf. Pielou 1991). No higher plants can survive in areas covered by ice. Since evidence of glaciation is found nearly everywhere in Fennoscandia and in the British Isles, the *tabula rasa* hypothesis, supposing that all the higher plants are post-glacial immigrants from south of the ice sheet, is fascinating and old in Scandinavia. It found strong support when Nathorst (1892) discovered fossil remains of typical alpine plants in deposits formed during the deglaciation of South Scandinavia.[2]

Since then, many attempts have been made to explain the phyto-geographical picture of Northern Europe without any supposition of plant survival in unglaciated refuges during the Pleistocene glaciations (cf. Nordal 1987; Birks 1993; H. H. Birks 1994). Following a modern and more elaborate *tabula rasa* hypothesis, the flora of Northern Europe is comprised of immigrants from unglaciated areas south, west and east of the North European ice sheets (H. H. Birks 1994).

The many Amphi-Atlantic species in Northern Europe, absent from the Central European mountain ranges, are generally supposed to have arrived from the west by long-distance dispersal (Brockmann-Jerosch 1923; Savile 1972, 1981; Heslop-Harrison 1973; Nordal 1985a) or gradually from the east as the ice melted, resulting in a once continuous, later disrupted, circumpolar distribution (Hultén 1955). Evolutionary rates are supposed to have been high enough to permit differentiation of many endemic species and subspecies in the course of the 18 000 years since the maximum glaciation (Nordal 1985b, 1987). Centricity, or the existence of particular areas with concentrations of endemic and disjunct taxa, could have resulted from special environmental conditions in or between the centres (Blytt 1876, Salisbury 1935, Bøcher 1951, Dahl 1951, Birks 1993). There is more and more evidence from finds of fossil seeds and fruits of the occurrence of centric and disjunct taxa during the last glaciation in areas south, west, and east of the main ice sheets (H. H. Birks 1994). Fossil finds of *Papaver* sect. *Scapiflora*, *Braya linearis*, *Cassiope hypnoides*, *Rhododendron lapponicum*, *Pedicularis hirsuta* and *C. tetragona* in areas

---

[2] The alternative *nunatak* hypothesis was firstly proposed by Sernander (1896) for Scandinavia. A corresponding hypothesis pertaining to Greenland had already been proposed by Warming (1888) and later similar hypotheses have been proposed for other North Atlantic regions (cf. Dahl 1989).

as far away as southern England, Ireland, Scotland, southern Sweden, western Norway, North Norway and northern Siberia led H. H. Birks (1994) to conclude:

> although there is no fossil evidence directly for or against the nunatak theory, the available fossil evidence indicates that it is not necessary to propose periglacial survival on nunataks or other ice-free refugia as an explanation of present mountain plant-distribution in Norway.

10   Anthropochorous plants

Plants that benefit from human activities and thereby increase their population size or geographical area are termed **synanthropic** or **hemerophilous**. This includes weeds in fields or gardens, also called **agrestals** or **segetals**, and ruderals that occupy areas where human activity has disturbed the natural vegetation in abandoned plots, along roads or railways, etc. Weeds are mostly annual, whereas ruderals are mostly perennial.

Phytogeographical problems of anthropochorous plants have been treated by Linkola (1916, 1921), Jessen & Lind (1923), Salisbury (1961), Berglund (1966a, b), Godwin (1975), Holzner & Numata (1982), Mucina *et al.* (1984), Willerding (1986), Di Castri *et al.* (1987), Kornek & Sukopp (1988) and Sukopp & Hejny (1990). The role of man in European vegetation history has been summarised by Behre (1988).

Before the advent of agriculture in the Neolithic period man lived as a hunter-gatherer and his impact on the flora was different from animals in only a few ways. He contributed to the dispersal of diaspores and no doubt encouraged growth of some species around his habitations as some species benefit from the addition of nitrogen and phosphorus to the soils. The use of fire may also have affected the vegetation, but we know little about this.

With the advent of agriculture, the effects became more important. The first agriculturalists used fire as a means to clear the vegetation. Trees were felled with stone axes and the slash thus produced was burnt. The previous vegetation was thus destroyed and the ash enriched the soils with plant nutrients. Seeds of wheat (*Triticum*) and barley (*Hordeum*) were sown in the ashes. A burned area could give good crops for a few years, but when the extra nutrients were gone, the native vegetation quickly invaded and the plot was abandoned while a new plot was prepared. This has been shown by palynological evidence and confirmed by an experimental felling of oaks with stone axes in Denmark and subsequent burning and sowing in the ashes (Iversen 1956, 1973). Weeds invaded the small fields and in pollen diagrams the first appearance of cereal grasses is followed by a rise of weed species, especially *Plantago lanceolata* and *P. major*, *Artemisia*, and Chenopodiaceae (Iversen 1973).

Along with early agriculture, animal husbandry was also introduced. Elm was especially favoured for collecting leaf-fodder. In the pollen diagrams, a rapid fall in the elm curve coincides more or less with the appearance of cereal grasses and weeds. In northern Norway, a period with animal husbandry preceded the arrival of cereals and weeds (Vorren 1979, 1986).

According to Godwin (1975, p. 471) the vegetation of the southern parts of the British Isles was dominated by forests, perhaps with shrubs along the seashore or on steep, dry and rocky sites. In the Neolithic, shrubs began to dominate and new shrub species appeared in the pollen diagrams (e.g. Peglar 1993).

At a much later stage, in the late Bronze Age in Scandinavia, the plough came into use as shown by rock carvings depicting an ox drawing a primitive plough, an ard. This permitted the formation and maintenance of permanent fields.

Anthropochorous plants are plants that have expanded following agriculture. With the advent of agriculture, man created new ecological habitats that many plants could invade. For a good crop, light, nutrients and an absence of competition are essential. The weeds compete with crop plants, and many techniques were developed to control weeds, not always with success.

In the fields some native species invaded, and these are called **apophytes**. These are species adapted to open habitats and a high nutrient content in the soil. One natural habitat for such species is along seashores where the proximity of the sea, strong winds, and salt spray all limit tree growth. On drift piles of seaweed along the shores, there are high quantities of available nitrogen and phosphorus. From such nitrophilous shore communities deep-rooted perennial species such as *Elytrigia* (*Agropyrum*) *repens*, *Cirsium arvense* and *Sonchus arvensis*, and many annuals such as *Sonchus asper* and *S. oleraceus* spread. Another nutrient-rich community is the tall-herb meadows from which species like *Tussilago farfara* and *Anthriscus sylvestris* probably came. Several of the weeds came from dry, steppe-like communities or from communities on shallow soils where drought prevented closed forest growth.

Species not previously present also came as weeds with seeds of cultivated plants. From the first agriculture we have evidence of the immigration of weeds. Several weed species can be identified as pollen. With early agriculture, pollen of *Plantago lanceolata*, *P. major* and *Centaurea cyanus* appear in the pollen diagrams together with the pollen

150

of crop plants and evidence of deforestation (Iversen 1949, 1973). Sometimes the food stores were burnt and some of the crop became carbonised. The carbonised remains are not easily broken down by micro-organisms and are preserved. In such samples weed species can also be identified. Additional evidence is provided by seeds being incorporated in the clay used to make pottery. When the clay was burnt the seeds were oxidised but left a print sufficiently detailed to permit the identification of the seed.

With the further diversification of agricultural crops new weeds were introduced. Still later, as a result of horticulture, especially in medieval monasteries, medicinal plants, plants used as spices, and some ornamental plants were introduced and became naturalised. Military campaigns also brought in weeds, for example *Bunias orientalis* during the Napoleonic wars. After the great explorations new species were introduced from, for example, America.

It is by no means easy to decide whether a plant is indigenous, that is present without the help of man, or if it is an invader absent from the area before agriculture or other ecological disturbances due to human activities (Webb 1985). The following criteria can be applied.

1 The easiest criterion is when we can observe from historical evidence, if not necessarily the actual introduction, the subsequent phase of expansion.
2 Plants whose existence today is dependent upon ecological conditions that exist only as a result of human activities such as cultivated fields, roadbanks, waste-heaps, etc. in the immediate surroundings of human habitations. They do not enter what is generally regarded as ecosystems or plant communities that are independent of man. It is thought that if man disappeared from the globe such species would be competed out of existence by forest regrowth and probably therefore had no natural ecological niche before the advent of agriculture. However, this is not an infallible criterion for two reasons:
(a) Some species, although not many, for which we have been able actually to follow the expansion, have invaded communities which, on all criteria, are natural communities where they compete on equal terms with the original flora. Examples include *Senecio viscosus* which came to Norway in about 1800 but now enters seashore communities that are as natural as any that can be found. Species of the *Epilobium adenocaulon* group coming from North America are now present in wet meadow communities that are also considered natural.

(b)   A number of species which on the basis of criterion 2 might be considered introduced have been found as fossil seeds or pollen in remains dating back to a time before agriculture was introduced to the area. Many have been found as fossils in glacial or late-glacial deposits as components of the open communities that occurred during these stages. Examples are *Centaurea cyanus* and different *Polygonum* species (Iversen 1949; Godwin 1975; Willerding 1986). It is possible that these open-habitat plants became extinct during the early Holocene, and were re-introduced with agriculture. However, some species have been identified as fossils occurring during the Holocene before the Neolithic or during previous interglacials when there was no agriculture. They were probably present somewhere in open communities, perhaps on drift along seashores or along streams, or on dry rock ledges or shallow soils.

A useful criterion for deciding whether a plant is anthropochorous or is an apophyte is when the species is represented by different races. Sometimes there is a native race and another which is naturalised. Examples include *Asparagus officinalis* with ssp. *officinalis* which is introduced and ssp. *prostratus* which is probably native as it grows on grassy sea-cliffs, and carrot (*Daucus carota*) which is represented by the introduced ssp. *sativa* and the native ssp. *gummifer* which grows naturally on sea-cliffs and dunes. *Crepis tectorum* is represented by an endemic race in Öland. *Erysimum cheiranthoides* is represented by a northern ssp. *altum* in Scandinavia that differs from the common weedy ssp. *cheiranthoides*. *Trifolium rubrum* in Scandinavia is mostly an escape from clover cultivation, but on sand-dunes along the Norwegian coast native races are found that are different from the cultivated plants.

Anthropochorous plants are also recruited from horticulture. This applies to some useful species and it is almost impossible to decide, for example, whether *Malus sylvestris* and *Prunus avium* belong to the native flora or are escapes from gardens. Apple seeds have been found from early Neolithic times (Hjelmqvist 1955) and are present in the Oseberg grave in Norway from the time of the Vikings. Some species that are found naturalised in the vicinity of medieval monasteries are suspected to have been introduced.

It is customary to divide the anthropochorous plants into **archaeophytes** which invaded before 1500 and **neophytes** which came later. There are several reasons for this. After medieval times, written sources become more helpful. With the discovery of the other continents new species became introduced into Europe. Often they came from botanic gardens where exotic plants were grown. The typical pattern is that after having been grown for a number of years the cultivated species spread to the surroundings and became naturalised. Perhaps several generations were needed to develop ecotypes that were able to survive under the new environmental conditions. Diaspores were also introduced around ports and became naturalised. Species such as *Robinia pseudacacia, Phytolacca americana, Mimulus guttatus* and *Epilobium* spp. of the *adenocaulon* group, came from North America. *Galinsoga parviflora* came from South America and, more recently, the Austral regions have contributed *Fuchsia magellanica* and *Epilobium brunnescens.*

In addition there are **ephemerals**. They are unable to reproduce themselves under the new conditions, and require repeated introductions to remain members of the flora. In local floras they are often printed in small text.

The proportion of introduced species is very different for different taxonomic groups. There are few, if any, naturalised species among the pteridophytes, the orchids, and the Cyperaceae. Families with many naturalised species include Polygonaceae, Chenopodiaceae, Brassicaceae, Asteraceae and Poaceae. A high proportion are annuals.

I have made a comparison of the number of species in the different categories of native species, archaeophytes and neophytes in the flora of the British Isles (based on Clapham *et al.* 1987) and the flora of Norway, Sweden and Finland (based on Lid 1985). All species are included except for the polymorphic, apomictic groups of *Rubus, Taraxacum* and *Hieracium.* Ephemerals are not included. Major sources of information include Linkola (1916, 1921), Jessen & Lind (1923), Iversen (1949), Lagerberg *et al.* (1950–8), Hjelmquist (1955), Fægri (1958–60), Salisbury (1961), Meusel *et al.* (1965, 1978), Berglund (1966a, b), Godwin (1975), Willerding (1986), Svensson & Wigren (1986), and Kornas (1987). The comparison is based on the following features.

1 If the taxon involved is restricted today to human-influenced habitats, and in the absence of any fossil evidence to the contrary, the taxon is considered an **immigrant**.
2 The taxon is considered an **archaeophyte** if it is a field weed or a useful plant in some way, as supported by fossil or literary evidence. Species that occur in glacial and late-glacial deposits but are absent from

interglacial or early Holocene deposits are considered to have been re-introduced by agriculture. Furthermore, species considered useful for medical purposes or as spices and that might have been grown in medieval gardens are considered archaeophytes.

3 Other species are considered **neophytes**. This includes all species where written sources or specimens in herbaria tell us when the species first appeared and, furthermore, species coming from distant countries in the New World, the Austral region, or the Far East. Also included are naturalised ornamental species unless there is evidence to the contrary.

The results are given in Table 6 which also includes data from Berlin based on Kovarik (1990).

Table 6. *Comparison of numbers of native and different anthropochorous species in Fennoscandia, the British Isles and Berlin*

	Native	Archaeophytes	Neophytes
Fennoscandia	1455	161	122
British Isles	1426	152	356
Berlin	839	167	426

The number of native species is nearly the same in Fennoscandia and the British Isles despite the fact that the area of Finland, Norway and Sweden taken together is 1 115 000 km² whereas the area of the British Isles is only 312 000 km². The species density is higher in the British Isles than in Fennoscandia. The number of archaeophytes is nearly the same, while the number of neophytes is much higher in Britain and Berlin than in Fennoscandia. Of a total flora of 2995 species, native or naturalised, in the former West Germany, Kornek & Sukopp (1988, p. 31) consider 267 or 10% as neophytes.

With the start of the Industrial Revolution about 200 years ago, agriculture changed with important consequences for the weed flora. Most important was the invention of artificial fertilisers. Before their introduction, farmers depended on transport of plant nutrients from the surrounding grasslands and forests to fertilise the fields to maintain their productivity. This took place by the grazing of cattle and sheep outside the fenced fields during the day and bringing them into the barns at night. Hay was also cut and trees were pollarded and the branches fed to the animals. The animals droppings were used as fertilisers. Areas for grazing, haymaking and pollarding about five times the size of the fields were needed to supply the plant nutrients required to sustain the agricultural production in the fields. With the introduction of artificial fertilisers, the areas previously used for grassland and forest were converted to agricultural fields and managed forests for timber production. The rural landscape portrayed on pictures from 150 years ago is rapidly disappearing. In the modern agro-industrial landscape, there are simply many extensive fields and managed forests. It is now an important task for conservation to preserve some examples of the old rural landscape.

With modern technology the threat of most weeds to crops has been eliminated and many have become rare and threatened by extinction. Better methods for cleansing the seeds of the crops before sowing have been an important cause of this decline. Thus *Apera spica-venti* which was listed as common in Norway has now become very rare. Further, the use of herbicides as well as heavy dressing of nitrogen compounds in modern agriculture has eliminated many weed species that were previously considered serious pests (Willerding 1986; Wilson 1991). The colourful fields with abundant poppies and cornflowers have largely gone. Modern fields are thus very dull from a botanical point of view. The most important weeds remaining are geophytes such as *Cirsium arvense*, *Elymus repens* and *Sonchus arvensis*. Of the annual weeds, only *Avena fatua* remains a pest in oat fields. Among the plants threatened by extinction in the Red Lists of many countries, there are many annual weeds, for example annuals associated with flax agriculture.

Species influenced by the spillage of dung in farmyards form another group of anthropophytes threatened by extinction. These are highly nitrophilic species belonging in the communities of Chenopodietea (Oberdorfer 1983). In modern animal husbandry such spillage is practically eliminated. This group includes several Brassicaceae such as *Sisymbrium* spp., and Boraginaceae such as *Asperugo procumbens* and *Lappula deflexa*.

Some of the species of the grazing areas and the cut hay-meadows are also about to disappear. In the dry Festuco-Brometea communities of Europe (Oberdorfer 1978), with their flora of numerous orchids, species such as *Orchis* spp., *Ophrys* spp., *Anacamptis pyramidalis*, *Aceras anthropoporum* and *Himantoglossum hircinum* are threatened. Also the famous '*lövängar*' in Gotland in Sweden, as well as species of *Botrychium* and *Gentianella* are threatened. Ploughing and re-seeding of old grasslands provide further threats for many traditional grassland species (Ratcliffe 1984).

Of the 63 species listed as extinct in the former West Germany (Kornek

& Sukopp 1988), 18 are weeds and 17 are species of dry or moist meadows resulting from the old agricultural practices. Thus agricultural change has been the most serious threat to the diversity of the present German flora. Programmes have therefore been proposed to conserve some areas using traditional agricultural methods in an attempt to preserve the agrestal weed species (Hilbig 1982) or to permit weeds to persist in some areas set aside for such purposes within the agricultural landscape (Wilson 1991). There are also programmes for preserving Mesobromion communities with their rich flora of orchids as well as other communities affected by old agricultural methods. Hillier *et al.* (1990) discuss a range of approaches to the preservation and conservation of Mesobromion calcareous grassland communities.

Calculation of climatic parameters for comparison with plant distributional data

Several climatic parameters have been calculated for each of the the 50 × 50 km grid squares of *Atlas Florae Europaeae* (O. Skre, unpublished data). The main source of the primary meteorological information is Anonymous (1973). In some instances where this is insufficient, for instance for Svalbard, information has been taken from other sources. The highest or lowest altitudes for each of the squares have been taken from national atlases. They are considered to be accurate to within about 10 m.

The following parameters have been used as a basis for the comparative maps (Figs. 12–17, 30–38, 42–50 and 54–63).

Winter temperatures (for comparisons with the atlantic and the boreal distribution patterns)

This is the mean temperature of the coldest month for each station. One problem is that in oceanic regions February is the coldest month, whereas in continental stations it is January, and in intermediate areas it is either January or February. To overcome this, we have used the normals of the 5 coldest months at each station. By means of least-square methods a parabola was fitted to the observations. The mean temperature of the coldest month is then the minimum of the parabola. It is always near the temperature of the coldest month.

By comparison with records from neighbouring stations at different altitudes, it is evident that the altitudinal temperature gradient differs as a function of continentality. In Portugal the gradient is about 0.6 °C per 100 m altitudinal difference, whereas in northern Russia it is zero. We have used a gradient taken from a linear regression of annual amplitude (difference between coldest and warmest month) with fixed points: Portugal 0.6 °C and northern Russia 0.0 °C per 100 m altitude difference. An isotherm map for sea level (W0) was constructed and subsequently the temperatures for the highest (Wh) and lowest (Wl) points in each square were calculated. The temperatures for the lowest altitudes were used for comparison with distributional patterns of atlantic or oceanic species, those for the highest altitudes for comparison with boreal distributional patterns.

The respiration sums (for comparison with the thermophilic distribution patterns)

The justification of using the accumulated annual respiration or respiration sums

(Re) and the R-values ($=\ln$ Re) has been discussed on p. 62. Here the procedure for the calculations is explained.

It builds on a model where the parameters are based on the physiological properties of *Picea abies* (Dahl & Mork 1959; Skre 1971, 1972, 1979a), but it can also be calculated for different parameters. It is assumed that respiration within the temperature ranges considered here follows the Arrhenius equation

$$\mathrm{d}\ln \mathrm{Re}/\mathrm{d}T = -q/R\,T^2 \qquad (1)$$

with an activation energy, q, of 70 kJ/mol for spruce corresponding to a Q_{10} ratio of 2.65. R is the gas constant ($= 8.314$ kJ/mol) and T is the temperature in degrees Kelvin ($=$ °C $+ 273$). When we want to calculate the respiration sum throughout the growing season, we define as one unit the amount of respiration at 10 °C for one month. Thus a respiration sum of 2.65 corresponds to 2.65 months at 10 °C or 1 month at 20 °C.

However, not all of this respiration is available for growth processes. Some is used for maintenance and transport. This basal respiration corresponds to the respiration at a constant temperature of 2.8 °C in spruce. Only the respiration above this value is considered available for growth processes.

The duration of the growing season is defined as the period from when the temperature rises above 2.8 °C in spring until it falls below this value in autumn.

Thus the growth respiration sum, Re, is the difference between the total respiration Re(tot) and what is required for maintenance and transport Re(mt).

We need to estimate the mean and the moments around the mean. The total respiration sum in a month is, according to Skre (1972, p. 17):

$$\mathrm{Re(tot)} = \exp(-q/R(1/M - 1/283) \cdot (1 + \mu^2 q^2/(2R^2 M^4) + \mu^3 q^3/(6R^3 M^6) + ...) \qquad (2)$$

This forms a highly convergent series where the later terms can be ignored. q is the activation energy, M is the mean in degrees Kelvin, R is the gas constant, μ^2 is the second moment around the mean, and μ^3 is the third moment around the mean.

Estimates of the second and third moments around the mean can be calculated from standard data in meteorological yearbooks using the absolute maxima and minima, the mean daily maxima and minima, and the mean daily temperatures as predictors. The empirical relations have been based on data from a number of northwestern European stations where hourly data are available (Skre 1971). Skre found that the standard deviation of the temperature, s, could be estimated by the equation

$$s = 0.31\, tx - 0.22\, tn - 0.29\, \underline{tx} + 0.17\, \underline{tn} + 0.034$$

where tx is the average monthly maximum, tn is the average monthly minimum, $\underline{tx}$ is the mean daily maximum, and $\underline{tn}$ is the mean daily minimum. The μ^2 in (2) is the square of s.

For calculation of μ^3 Skre (1972) found that it could be estimated from the equation

$$\mu^3/\mu^2 = 0.82\ tx + 0.76\ tn - 0.71\ \underline{tx} - 0.81\ \underline{tn} - 1.5072 \qquad (3)$$

Now, by inserting estimates of the second and third moment around the mean in (2) the total respiration is calculated for each month and for the whole growing season.

These calculations have been made for each station. But since the stations are located at different altitudinal levels we must correct for altitude in order to compare stations. Hence we have to find a correction for altitude.

The mean temperature in (1) is a function of altitude. We assume that the temperature gradient during the summer months is near to 0.6 °C per 100 m altitudinal difference. If the mean temperature at sea level is T0 in the station T, and the altitude of the station is h we obtain

T = T0 − h 0.006 or

dT = − dh 0.006

Inserting in (1) and integrating we get

$$\ln\ (Re/Re0) = -\ 50.35\ h/T0^2 \qquad (4)$$

where h is measured in metres and T0 is the estimated mean temperature at sea level. This is Re(tot).

We can see that the total respiration sums vary logarithmically with altitude. Skre (1972) found empirically that in Scandinavia growth respiration varied logarithmically with altitude according to the following equation:

$$\ln\ (Re/Re0) = -\ 0.001\ h$$

This formula pertains to the growth respiration. Using (3) with a mean temperature during the growth season of 281 (West Norway) we get a coefficient of 0.00065.

An advantage of using natural logarithms is that it permits a direct comparison between latitudinal and altitudinal variation. Let us calculate the expected difference in altitudinal limits for plants in the Alps and in South Norway. The logarithm of the total respiration sum at sea level in Southeast Norway amounts to about 2.40. The corresponding value for the Alps is close to 3.00 while the mean temperature at sea level is 286 °K. Inserting these values in (4) and solving for h gives an estimate for $h = 975$ m. We have found an altitudinal difference of about 1000 m (which also compares with the climatic timber-line, p. 55).

In the maps we have used the ln transformations of the respiration sum data. These new measures are called R-values. Thus

R = ln Re

This has been used for total respiration, for growth respiration, and for respiration sums calculated for different altitudes in the landscape.

We have calculated total respiration sums and growth respiration sums for all stations included in Anonymous (1973). We have also calculated the duration of the growing season at sea level, the mean temperature of the growing season at sea level, and R-values both for total and for growth respiration at sea level.

For comparison with plant distribution maps, for example the *Atlas Florae Europaeae* maps, the altitude of the landscape must be taken into account. From the above formulae we have calculated the respiration sums, expressed as R-values for the lowermost station within each of the 50 × 50 km squares used in the Atlas. These are compared with the distribution, and especially the northern limits, of thermophilic plants.

In order to obtain the growth respiration we must substract the basal respiration over the growing season. In spring and autumn the duration of the growing season in northwestern Europe decreases with 8 days per 100 m altitudinal difference (Skre 1979a). We can therefore calculate the duration of the growing season at sea level, D0, when the duration at station level is D

$$D0 = D + 0.08\ h \tag{4}$$

Substracting the daily basal respiration multiplied by the duration gives the growth respiration

$$Re(growth) = Re(tot) - 0.0145\ D \tag{5}$$

The daily basal respiration corresponding to 2.8 °C is 0.0145 units when the respiration at 30 days at 10 °C is taken as unity.

Maximum summer temperature (for comparison with arctic and alpine distribution patterns)

The measure used is the mean of the annual maxima for the meteorological stations. The altitudinal gradient has been taken to be 0.7 °C per 100 m difference. First, the maximum temperature at sea level has been calculated and subsequently the temperature at the highest points in the 50 × 50 km squares. These values are compared with the distribution of arctic and alpine species (Dahl 1951).

Drought stress

As a first measure the total amount of precipitation falling in summer has been calculated, summer being defined as the months with a mean temperature of 5 °C or higher. The summit stations have been avoided.

Drought stress for stenohydric plants arises when evaporation exceeds the available amount of water stored in the soil or falling as rain. According to Gates (1980) the evaporation E (mm H_2O) from a grass turf supplied with enough water is

$$E = 0.32 \, (e(t) - e(a))(1 + 0.18 \, V)$$

where $e(t)$ is the saturation pressure (millibars) in air at temperature t and $e(a)$ is the partial pressure of water in the air. V is the wind velocity at 2 m altitude.

The following parameter is used here as a measure of drought stress

$$D_r = E - P$$

where P is the precipitation.

Disregarding the wind factor, D_r can be calculated from data in Anonymous (1973) which gives average daily maximum and minimum temperatures for each month as well as relative humidity for midday and early morning. From the maximum temperature and the humidity at midday, a midday saturation deficit can be calculated, and in the same way, a saturation deficit for early morning can be derived. The daily average value is taken as the mean of these two values.

Such values can be calculated for each month and the data used here are for the month in each station where E – P reaches its maximum. In northwestern Europe this is usually in early summer, May or June, but in the east it is later in summer, July, and in the Mediterranean it is often in August.

The Northern European species of *Flora Europaea* with indications of their status and climatic correlations

BI is British Isles, Fe is Fennoscandia. Information on status in the two areas is based on available evidence – i is indigenous, n is naturalised, e is ephemeral, * is endemic to the area.

Wl is the mean temperature (°C) of the coldest month calculated for the lowest altitudes in the *Atlas Florae Europaeae* squares – to be compared with the distribution of atlantic species.

R is the R-value calculated for the lowest altitudes in the squares – to be compared with the distribution of thermophilic plants.

Wh is the mean temperature (°C) of the coldest month calculated for the highest point in each square – to be compared with the distribution of boreal species.

Tmax is the mean annual maximum temperature (°C) calculated for the highest point in each square – to be compared with the distribution of montane and alpine species.

The sequence of families and genera is according to *Flora Europaea* based on D. M. Moore: *Check-list and chromosome index. Flora Europaea*, 1982. The nomenclature is mainly according to the same source, but modified according to Clapham *et al.* (1987) and Lid (1985).

Although all species occurring in Northern Europe according to *Flora Europaea* are listed in this appendix, correlations between their distribution and the four climatic parameteres are only presented for those species where the correlations are realistic and ecologically plausible.

Species	Status		Climatic parameters			
	BI	Fe	Wl	R	Wh	Tmax
PTERIDOPHYTA						
LYCOPODIACEAE						
Huperzia selago	i	i	0.0	0		+32
Lycopodiella inundata	i	i	1.6			
Lycopodium annotinum	i	i	0.5	−1		+33
L. clavatum	i	i	0.6			+33
Diphasiastrum alpinum	i	i	0.4			+27
D. complanatum ssp. *complanatum*	i	i				+33
ssp. *montellii*		i				−11
D. chamaecyparissus		i	1.9			
SELAGINELLACEAE						
Selaginella selaginoides	i	i			0.4	−1
ISOETACEAE						
Isoetes lacustris	i	i				
I. echinospora	i	i				
I. histrix	i		+6			
EQUISETACEAE						
Equisetum hyemale	i	i	0.8			
E. ramosissimum	i		2.2			
E. variegatum	i	i	0.5	+1		
E. scirpoides		i		−4		
E. fluviatile	i	i				
E. palustre	i	i				
E. sylvaticum	i	i	0.6			
E. pratense	i	i	0.5	0		
E. arvense	i	i				
E. telmateia	i		2.0			
OPHIOGLOSSACEAE						
Ophioglossum lusitanicum	i		+6			
O. azoricum	i					

Species	Status		Climatic parameters			
	BI	Fe	Wl	R	Wh	Tmax
O. vulgatum	i	i	1.6			
Botrychium simplex		i				
B. lunaria	i	i	0.5	+2		
B. boreale		i		−5		
B. matricariifolium		i	1.6	−1		
B. lanceolatum		i		−8		
B. multifidum		i		−1		
B. virginianum		i	1.7	−4		

OSMUNDACEAE

Osmunda regalis	i	i	−4			

ADIANTACEAE

Anogramma leptophylla	i		+5			
Adiantum capillus-veneris	i		+4			
Cryptogramma crispa	i	i	0.4			+27

HYPOLEPIDACEAE

Pteridium aquilinum	i	i	1.4			

HYMENOPHYLLACEAE

Hymenophyllum wilsonii	i	i				
H. tunbrigense	i		+4			
Trichomanes speciosum	i					

THELYPTERIDIACEAE

Thelypteris limbosperma	i	i	−3			+32
T. palustris	i	i	−11	1.4		
T. phegopteris	i	i	−12	0.6	+1	+33

ASPLENIACEAE

Asplenium marinum	i	i	+3			
A. petrarchae			3.0			
A. trichomanes	i	i	1.6			

Species	Status		Climatic parameters			
	BI	Fe	Wl	R	Wh	Tmax
A. viride	i	i		0.6		+31
A. billotii	i			+1		
A. adiantum-nigrum	i	i	−2			
A. onopteris		i	+4			
A. septentrionale	i	i	1.4	0		
A. ruta-muraria	i	i	1.4			
Ceterach officinarum	i	i	0			
Phyllitis scolopendrium	i	i	−2			
ATHYRIACEAE						
Athyrium filix-femina	i	i				
A. distentifolium	i	i	0.4	−1		+27
A. flexile*	i					
Diplazium sibiricum		i	1.3	−9		
Cystopteris fragilis	i	i	0.8			
C. dickieana	i	i				
C. montana	i	i			−6	
C. sudetica		i	1.9			
Woodsia ilvensis	i	i				
W. alpina	i	i			−6?	
W. glabella		i			−11	
Matteuccia struthiopteris	n	i				+31
Gymnocarpium dryopteris	i	i	0.6			+32
G. robertianum	i	i	1.3			+33
ASPIDIACEAE						
Polystichum lonchitis	i	i	0.6			+30
P. aculeatum	i	i	−4			
P. setiferum	i		0			
P. braunii		i	1.6	−3		+31
Dryopteris filix-mas	i	i	0.9			
D. affinis	i	i	−2			
D. abbreviata	i					
D. villarii	i					

Species	Status		Climatic parameters			
	BI	Fe	Wl	R	Wh	Tmax
D. cristata	i	i		1.9 (1.7 in Ru, NF)		
D. carthusiana	i	i				
D. dilatata	i			1.8		
D. expansa	i	i		0.5		+30
D. aemula	i		+4			

BLECHNACEAE

Blechnum spicant	i	i				

POLYPODIACEAE

Polypodium australe	i		+2			
P. vulgare s.l.	i	i		0.8		
P. interjectum	i	i?				

MARSILEACEAE

Pilularia globulifera	i	i				

GYMNOSPERMAE

PINACEAE

Abies sibirica	e	n			−9.5	
A. alba	e	n			0	
Picea abies ssp. *abies*	i	i	1.1	−2		+33
ssp. *obovata*		i			−10	
Larix sibirica	n	n			−10	
Pinus sylvestris	i	i	1.1	−1		+33
P. mugo + *P. uncinata*	e	e				+30
P. contorta	n	e				
P. nigra	n					
P. sibirica	n	n			−12	
P. cembra	n	n				+29

CUPRESSACEAE

Juniperus communis ssp. *communis*	i	i				
ssp. *nana*	i	i	0.4	0		+29

Species	Status		Climatic parameters			
	BI	Fe	Wl	R	Wh	Tmax
TAXACEAE						
Taxus baccata	i	i	−4			
ANGIOSPERMAE						
SALICACAE						
Salix pentandra	i	i	1.2	+1		
S. alba	i	n	1.9			
S. triandra	i	i	1.4			
S. reticulata	i	i	0.3		+24	
S. herbacea	i	i	0.0		+25	
S. polaris		i	0.2		+24	
S. nummularia		i?				
S. myrsinites	i	i	0.5			
S. glauca		i	0.4		+30	
S. lanata	i	i	0.4			
S. phylicifolia	i	i	0.5	−4?	+29–30	
*S. hibernica**	i?					
S. myrsinifolia ssp. *myrsinifolia*	i	i	0.9	−1	+33	
ssp. *borealis*		i				
S. jeniseiensis				−12		
S. atrocinerea	i					
S. cinerea	i	i	1.5	+2		
S. aurita	i	i	1.5			
S. caprea	i	i	1.0	+2		
S. starkeana ssp. *starkeana*		i		−3		
ssp. *cinerascens*		i		−8		
S. myrtilloides		i		−3	+32	
S. repens	i		1.6			
S. rosmarinifolia		i		−1		
S. arbuscula	i	i				
S. hastata		i				
S. lapponum	i	i	0.5	−3	+31	
S. purpurea	i	n	2.0			
S. daphnoides	n	i				
S. viminalis	i?	n				
S. fragilis	i?	n				

Species	Status		Climatic parameters			
	BI	Fe	Wl	R	Wh	Tmax
Populus alba	n	e	2.1			
P. tremula	i	i	1.0			
P. nigra	i	e	2.0			
P. balsamifera		e				
P. trichocarpa	e					

MYRICACEAE

Myrica gale	i	i				

BETULACEAE

Betula pendula	i	i		1.3		
B. pubescens	i	i		1.0		
B. nana	i	i		0.4	−2	+30
Alnus glutinosa	i	i	1.5			
A. incana ssp. *incana*	n	i		1.0	−2	
ssp. *kolaensis**		i			−10	

CORYLACEAE

Carpinus betulus	i	e				
Corylus avellana	i	i	−9?	1.6		

FAGACEAE

Fagus sylvatica	i	i	−4			
Castanea sativa	n	e	+1			
Quercus ilex	n		+4			
Q. cerris	n					
Q. petraea	i	i	−4			
Q. robur	i	i		1.8		
Q. pubescens			−1	2.3		

ULMACEAE

Ulmus glabra	i	i		1.6	+2	
U. minor	n	i		2.0		
U. laevis		i		2.0		

Species	Status		Climatic parameters			
	BI	Fe	Wl	R	Wh	Tmax
CANNABACEAE						
Humulus lupulus	i	i		1.7		
URTICACEAE						
Urtica dioica ssp. *dioica*	i	i		0.9		
ssp. *sondenii*		i			−11	
U. urens	i	n		1.1		
Parietaria judaica	i	e	−1			
P. lusitanica		e		2.8		
SANTALACEAE						
Thesium alpinum		i				+31
T. humifusum	i	i				
LORANTHACEAE						
Viscum album	i	i	−8	2.0		
ARISTOLOCHIACEAE						
Asarum europaeum	i?	i			+1	
Aristolochia clematitis	n	n		2.1		
POLYGONACEAE						
Koenigia islandica	i	i		0.4		+23
Polygonum maritimum	i					
P. raii	i	i				
P. oxyspermum *		i				
P. aviculare	i	i				
P. arenastrum	i	i?				
P. minus	i	i		1.7		
P. mite	i	i		1.9		
P. hydropiper	i	i		1.7		
P. persicaria	i	i?		1.4		
P. lapathifolium	i	i?				
P. amphibium	i	i		1.1		
P. bistorta	i	n				

Species	Status		Climatic parameters			
	BI	Fe	Wl	R	Wh	Tmax
P. viviparum	i	i			−2	+29
P. polystachyum	n		+1			
Fallopia convolvulus	i	i?				
F. dumetorum	i	i				
Reynoutria japonica	n	e	−5			
R. sachalinensis	n	e	−4			
Oxyria digyna	i	i				+27
Rheum raponticum		n				
Rumex acetosella	i	i				
R. graminifolius		i				
R. alpestris		i	0.4	−10		+26
R. acetosa ssp. *acetosa*	i	i				
ssp. *serpentinicola**		i				
R. thyrsiflorus		n				
R. alpinus	n					+28
R. aquaticus	i	i			−1	
R. pseudonatronatus		i				
R. longifolius	i	i?			0	
R. hydrolapathum	i	i	1.9			
R. crispus	i	i	1.2			
R. conglomeratus	i	e				
R. sanguineus	i	i	2.0			
R. rupestris	i					
R. pulcher	i	e				
R. obtusifolius	i	n	1.5			
R. maritimus	i	i	1.6			
R. palustris	i	e	2.0			

CHENOPODIACEAE

Beta vulgaris ssp. *maritima*	i	i				
Chenopodium bonus-henricus	n	n	1.8	0		
C. multifidum		e	2.8			
C. foliosum		e				
C. glaucum	i?	i?				
C. rubrum	i	e	1.8			

Species	Status		Climatic parameters			
	BI	Fe	Wl	R	Wh	Tmax
C. hybridum	n	e		1.9		
C. polyspermum	i?	i?		1.8		
C. vulvaria	i	i				
C. urbicum	i?	i		1.9		
C. murale	n	i				
C. ficifolium	i?	e		2.1		
C. album	i?	i?				
C. suecicum	n	n				
Atriplex laciniata	i	i				
A. littoralis	i	i				
A. patula	i	i?		1.0		
A. calotheca		i				
A. prostrata	i	i				
A. glabriuscula	i	i				
A. longipes	i	i				
A. praecox	i	i				
Halimione portulacoides	i	i				
H. pedunculata	i			1.9		
Salicornia europaea	i	i		1.3		
S. pusilla	i					
S. procumbens coll.	n	n				
S. ramosissima	i					
S. dolichostachya	i	i				
Suaeda vera	i			2.2		
S. maritima	i	i		1.5		
Salsola kali	i	i		1.9		

AMARANTHACEAE

Amaranthus retroflexus	e	e				
A. lividus		n				

AIZOACEAE

Carpobrotus edulis	n		+6			

PORTULACACEAE

Montia fontana ssp. *fontana*	i	i				

Species	Status		Climatic parameters			
	BI	Fe	Wl	R	Wh	Tmax
ssp. *variabilis*	i	i				
ssp. *amporitana*	i	i				
ssp. *chondrosperma*	i	i				
Claytonia perfoliata	n	e				
C. sibirica	n	n				
CARYOPHYLLACAE						
Arenaria balearica	n					
A. humifusa		i				+20
A. norvegica ssp. *norvegica*	i	i				+22
ssp. *anglica**	i					
A. pseudofrigida		i				
A. gothica		i				
A. ciliata ssp. *ciliata*	i					
A. serpyllifolia	i	i		1.4		
A. leptoclados	i	i				
Moehringia trinervia	i	i			+2	
M. lateriflora		i	–5			
Minuartia viscosa		i				
M. hybrida	i					
M. verna ssp. *verna*	i	e				+29
M. rubella		i				+21
M. stricta	i	i	0.5			+22
M. rossii		i				
M. biflora		i	0.2			+26
M. sedoides	i					+23
Honkenya peploides	i	i				
Stellaria nemorum ssp. *nemorum*	i	i		0.8	0	
ssp. *glochidisperma*	i	i		2.0		
S. media	i	i				
S. neglecta	i	i				
S. pallida	i	i				
S. holostea	i	i		1.4		
S. uliginosa	i	i		1.4		
S. palustris	i	i				
S. fennica		i		–10		

Species	Status		Climatic parameters			
	BI	Fe	Wl	R	Wh	Tmax
S. hebecalyx		i				
S. crassipes		i				
S. longifolia		i			−3	+31
S. calycantha		i		0.6	−10	
S. crassifolia		i			0	
S. humifusa		i				
S. graminea	i	i				
Holosteum umbellatum	i	i		2.0		
Cerastium cerastoides	i	i		0.2		+26
C. arvense	i	i			+1	
C. regelii		i			−10	
C. glabratum		i				+26
C. alpinum	i	i		0.2		+27
C. arcticum	i	i				+23
*C. edmondstonii**	i					
C. fontanum ssp. *vulgare*	i	i			+1	
ssp. *scandicum*		i				
*C. scoticum**	i					
C. brachypetalum	i	i	−3	2.0		
C. glomeratum	i	i				
C. semidecandrum	i	i	−6			
C. pumilum	i	i	−3	2.0		
C. diffusum	i	i				
Moenchia erecta	i					
Myosoton aquaticum	i	i		1.9		
Sagina nodosa	i	i				
S. nivalis		i		0.3		+22
S. caespitosa		i				+21
S. subulata	i	i	−2			
S. saginoides	i	i		0.4	−3	+27
S. procumbens	i	i				
S. apetala	i	e	−2			
S. maritima	i	i				
Scleranthus perennis	i	i		1.8		
S. annuus	i	i		1.7		
Herniaria glabra	i	i		1.8		

Species	Status		Climatic parameters			
	BI	Fe	Wl	R	Wh	Tmax
H. ciliolata	i	i	+7			
Polycarpon tetraphyllum	i		+2			
Spergula arvensis	i	i				
S. morisonii		i		1.8		
S. rupicola	i		+5			
Spergularia maritima	i	i				
S. salina	i	i				
S. rubra	i	i		1.5		
S. bocconei	i		+6			
Corrigiola littoralis	i	e				
Lychnis flos-cuculi	i	i		1.2		
L. viscaria	i	i		1.4	0	
L. alpina	i	i		0.4	−2	
Agrostemma githago	i	i		1.5		
Silene nutans	i	i		1.6		
S. viscosa		i		1.9		
S. tatarica		i				
*S. uralensisssp. apetala**		i		0.3		+23
S. furcatassp. angustiflora		i			−11	
S. otites		i		2.0		
S. italica	n					
S. vulgaris	i	i		1.1		
S. uniflora	i	i				
S. acaulis	i	i		0.1		+25
S. rupestris		i			−2	+29
S. noctiflora	i	n		1.9		
S. latifolia	i	i		1.5		
S. dioica	i	i		0.4		
S. dichotoma		n				
S. gallica	i	e				
S. conica	i	n				
S. conoidea	e	e				
Gypsophila fastigiata		i				
G. muralis		i		1.8		
Saponaria officinalis	n	n		2.1		

Species	Status		Climatic parameters			
	BI	Fe	Wl	R	Wh	Tmax
S. ocymoides		e				+30
Petrorhagia prolifera		i	−2			
P. nanteuilii	i		+6			
Dianthus barbatus		n				
D. plumarius	n	e				
D. gallicus	n					
D. gratianopolitanus	i					
*D. arenariusssp. arenarius**		i				
ssp. *borussicus*		i				
D. superbus		i				
D. deltoides	i	i		1.6	+1	+33
D. armeria	i	i		2.0		
NYMPHAEACEAE						
Nymphaea alba	i	i				
N. candida		i		1.2		+34
N. tetragona		i		1.4	−6	
Nuphar lutea	i	i		1.1		
N. pumilum	i	i		1.1	−1	+32
CERATOPHYLLACEAE						
Ceratophyllum demersum	i	i				
C. submersum	i	i		2.0		
RANUNCULACEAE						
Helleborus foetidus	i		0			
H. viridis	i					
Eranthis hyemalis	n	e				
Nigella arvensis	e			2.2		
Trollius europaeus	i	i			−1	+32
Actaea spicata	i	i		1.2	+1	+33
A. erythrocarpa		i		1.1	−9	
Caltha palustris	i	i				
Aconitum lycoctonum		i			−7	

Species	Status		Climatic parameters			
	BI	Fe	Wl	R	Wh	Tmax
A. napellum ssp. *napellum**	i	e				
Delphinium elatum		n			–6	
Consolida regalis	e	e				
C. ajacis	e	e				
Anemone nemorosa	i	i		1.4		
A. ranunculoides		i		1.4		
A. sylvestris		i				
Hepatica nobilis		i		1.6	+1	
Pulsatilla vernalis		i			–1	
P. pratensis		i		1.9		
P. vulgaris	i	i		1.9		
P. patens		i				
Clematis vitalba	i	i	–1			
C. recta	e	e		2.1		
C. sibirica		i			–9	
Adonis vernalis		i		2.0		
A. annua	n	e	+1			
Ranunculus polyanthemos		i		1.6		
R. repens	i	i				
R. acris ssp. *acris*	i	i		0.3		
ssp. *borealis*		i				
ssp. *pumilus*	i	i				
R. bulbosus	i	i		1.8		
R. sardous	i	i		2.0		
R. arvensis	i	i		2.0		
R. parviflorus	i		+3			
R. paludosus	i					
R. illyricus		i				
R. auricomus s.l.	i	i				
R. monophyllus		i				
R. pygmaeus		i		0.1		+24
R. nivalis		i				+23
R. sulphureus		i				+23
R. hyperboreus		i			–10	+27
R. sceleratus	i	i		1.6		

Species	Status		Climatic parameters			
	BI	Fe	Wl	R	Wh	Tmax
R. gmelinii		i			−10	
R. lapponicus		i				
R. pallasii		i			−11	
R. cymbalaria		n				
R. ficaria ssp. *bulbifer*	i	i				
ssp. *ficaria*	i					
R. platanifolius		i			−3	+28
R. glacialis		i		0.0		+22
R. parnassifolius						+24
R. flammula	i	i	−9	15		
ssp. *scoticus**	i					
ssp. *minimus**	i					
R. reptans	i	i				
R. lingua	i	i		1.8		
R. ophioglossifolius	i	i	0			
R. hederaceus	i	i				
R. omiophyllus	i					
R. tripartitus	i					
R. peltatus ssp. *peltatus*	i	i				
ssp. *baudotii*	i	i		1.7		
R. penicillatus	i			1.9		
R. aquatilis	i	i		1.9		
R. trichophyllus						
ssp. *trichophyllus*	i	i				
ssp. *eradicatus*		i				+28
R. circinatus	i	i		1.8		
R. fluitans	i	i		2.0		
Myosurus minimus	i	i		1.7		
Aquilegia vulgaris	i	n		2.0		
Thalictrum aquilegiifolium		i				
T. alpinum	i	i		0.3		+26
T. minus ssp. *minus*	i	i				
ssp. *kemense**		i			−9	
T. simplex		i		1.3		
T. flavum	i	i				

PAEONIACEAE

Paeonia anomala | −10

Species	Status		Climatic parameters			
	BI	Fe	Wl	R	Wh	Tmax
BERBERIDACEAE						
Berberis vulgaris	n	n				
PAPAVERACEAE						
Papaver somniferum	n	e				
P. rhoeas	i	n				
P. dubium	i?	n				
P. hybridum	n	e				
P. argemone	n	n				
P. lecoqii	i?	e				
P. atlanticum	n	e				
*P. radicatum**		i				
*P. lapponicum**		i				
Meconopsis cambrica	i	n				
Glaucium flavum	i	i				
Chelidonium majus	n	n				
FUMARIACEAE						
Corydalis claviculata	i	i	0			
C. solida	e	n				
C. bulbosa	e	i			0	
C. lutea	i	n			−1	
C. intermedia		i				
C. pumila		i			−1	
C. nobilis		n				
C. solida	n	i				+1
*C. gotlandica**		i				
*Fumaria occidentalis**	i					
F. capreolata	i	e				
*F. purpurea**	i					
F. bastardii	i		+4			
F. martinii	i		+5			
F. muralis	i	n				
F. borae		i				
F. densiflora	i	e				
F. officinalis	i	n				

Species	Status		Climatic parameters			
	BI	Fe	Wl	R	Wh	Tmax
F. vaillantii	i	n				
F. parviflora	n					
BRASSICACEAE						
Sisymbrium officinale	i	i		1.4		
S. irio	n	e				
S. loeselii	n	e				
S. volgense	n	n				
S. orientale	n	e				
S. altissimum	n	n				
S. strictissimum	n	e				
S. supinum		i				
S. austriacum		e				
Descurainia sophia	i	i				
Alliaria petiolata	i	i		1.7		
Eutrema edwardsii		i				
Arabidopsis thaliana	i	i				
*A. suecica**		i				
Braya linearis		i				+23
B. purpurascens		i				
Camelina sativa	n	n				
C. microcarpa		n				
C. alyssum		n				
Isatis tinctoria	i?	n				
Bunias orientalis	n	n				
B. erucago	n	e				
Erysimum cheiranthoides	i	i				
E. hieraciifolium		i				
Hesperis matronalis	n	n				
Cheiranthus cheiri	n			2.4		
Matthiola incana	i					
M. sinuata	i		+5			
Barbarea vulgaris	i	n				

Species	Status		Climatic parameters			
	BI	Fe	Wl	R	Wh	Tmax
B. stricta	i	i				
B. intermedia	n					
B. verna	n	e				
Rorippa austriaca	n	n		1.9		
R. sylvestris	n	n				
R. islandica	i	i				
R. amphibia	i	n				
R. palustris	i	i				
Armoracia rusticana	n	n				
Nasturtium officinale	i	i				
N. microphyllum	i	n				
Cardamine bulbifera	i	i	−6	1.8		
C. amara	i	i				
C. nymanii		i				
C. pratensis	i	i				
C. impatiens	i	i		1.8		
C. parviflora		i				
C. bellidifolia		i		0.1		+23
C. flexuosa	i	i	−6			
C. hirsuta	i	i	−6			
Cardaminopsis arenosa		i				
C. petraea	i	i				
Arabis glabra	i	i				
A. hirsuta	i	i		0.9		
*A. brownii**	i		+6			
A. turrita	n					
A. alpina	i	i		0.2		+26
A. caucasica	n	n				
A. planisiliqua		i				
A. pendula		e				
A. stricta	i					
Lunaria rediviva	e	i		1.9		
L. annua	n	n				
Alyssum alyssoides	n	n				
Berteroa incana	n	n				

Species	Status		Climatic parameters			
	BI	Fe	Wl	R	Wh	Tmax
Lobularia maritima	n	e				
Draba aizoides	i?	e				
D. alpina		i				+24
D. oxycarpa		i				
D. nivalis		i		0.4		+22
D. norvegica	i	i		0.3		+26
*D. cacuminum**		i				
D. fladnizensis		i				+22
D. lactea		i				+22
D. daurica		i				+27
D. incana	i	i				+29
D. muralis	i	i		1.9		
D. nemorosa	i	n				
D. crassifolia		i				+22
Erophila verna	i	i		1.7		
E. majuscula	i					
E. glabrescens	i					
Cochlearia danica	i	i				
C. officinalis	i	i				
C. pyrenaica	i					
*C. scotica**	i					
*C. micacea**	i					
C. fenestrata		i				
C. anglica	i	i				
Neslia paniculata		n				
Capsella bursa-pastoris	i	i				
Hornungia petraea	i	i	–4	1.9		
Teesdalia nudicaulis	i	i	–4	1.9		
Thlaspi arvense	n	n				
T. alpestre	i	n				
T. perfoliatum	i	n				
Iberis amara	i	e		2.2		
Lepidium campestre	i	i				
L. virginicum	e	n				
L. densiflorum	e	n				

Species	Status		Climatic parameters			
	BI	Fe	Wl	R	Wh	Tmax
L. neglectum		n				
L. heterophyllum	i	n				
L. ruderale	i	n				
L. latifolium	i	i		1.9		
L. graminifolium	n					
Cardaria draba	n	n				
Coronopus squamatus	i	n				
C. didymus	n	n				
Subularia aquatica	i	i				
Conringia orientalis	n	n				
Diplotaxis muralis	n	n				
D. tenuifolia	n	n		1.8		
Brassica oleracea	n	n				
B. napus	n	n				
B. rapa	n	n				
B. juncea	n	e				
B. fruticulosa	n					
B. nigra	i	n				
B. elongata	n	e				
Sinapis arvensis	i	n				
S. alba	n	n				
Erucastrum gallicum	n	e				
Rhynchosinapis cheiranthos	n	n		2.4		
*R. monensis**	i					
*R. wrightii**	i					
Hirschfeldia incana	n	e				
Cakile maritima	i	i				
Crambe maritima	i	i				
Raphanus raphanistrum	n	n				
R. maritimus	i					
R. sativus	n	e				

Species	Status		Climatic parameters			
	BI	Fe	Wl	R	Wh	Tmax
RESEDACEAE						
Reseda luteola	i	n				
R. lutea	i	n		1.8		
R. alba	n	e				
DROSERACEAE						
Drosera anglica	i	i		1.1		
D. rotundifolia	i	i		1.0		
D. intermedia	i	i		1.7		
CRASSULACEAE						
Crassula tillaea	i					
C. aquatica	n	i				
Umbilicus rupestris	i		+4			
Sempervivum tectorum	n	n				
Sedum telephium	i	i		1.7		
S. spurium	n	n				
S. reflexum	n	i	−3	1.8		
S. forsteranum	i		+1			
S. acre	i	i				
S. sexangulare	n	i			+1	
S. album	i	i				
S. anglicum	i	i	0			
S. dasyphyllum	n					
S. villosum	i	i				+24
S. annuum		i			−2	+28
S. rosea	i	i		0.1		+24
SAXIFRAGACEAE						
Saxifraga hieraciifolia		i				+22
S. nivalis	i	i	0.2		−4	+27
S. tenuis		i	0.1			
S. stellaris	i	i	0.2			+25
S. hirsuta	i		+7			
S. foliolosa		i				+24
S. spathularis	i		+5			

Species	Status		Climatic parameters			
	BI	Fe	Wl	R	Wh	Tmax
S. hirculus	i	i				
S. tridactylites	i	i		1.8		
*S. osloensis**		i				
S. adscendens		i			−4	+28
S. aizoides	i	i		0.5		+26
S. cespitosa	i	i		0.1	−5	+28
*S. hartii**	i					
S. rosacea	i	e				+23
S. hypnoides	i	i	+1			
S. granulata	i	i				
S. rivularis	i	i		0.1		+23
S. cernua	i	i		0.1		+26
S. oppositifolia	i	i		0.1		+26
S. cotyledon		i		0.9		+26
S. paniculata		i				
Chrysosplenium alternifolium	i	i		1.6		
C. oppositifolium	i	i	−1			
C. tetrandrum		i				

PARNASSIACEAE

Parnassia palustris	i	i		0.5	+2	

GROSSULARIACEAE

Ribes rubrum	n	n				
R. spicatum	i	i				
R. nigrum	n	i		1.7		
R. uva-crispi	i	n		1.8		
R. alpinum	i	i				

ROSACEAE

Spiraea salicifolia	n	n				
Filipendula vulgaris	i	i		1.8		
F. ulmaria	i	i		0.7		
Rubus chamaemorus	i	i		0.5	−1	
R. arcticus		i			−6	
R. saxatilis	i	i		0.7	0	
R. idaeus	i	i		0.8		

Species	Status		Climatic parameters			
	BI	Fe	Wl	R	Wh	Tmax
R. fruticosus coll.	i	i				
Rosa pimpinellifolia	i	i				
R. acicularis		i			−7	
R. majalis		i			−3	
R. rugosa	n	n				
R. virginiana	n					
R. canina	i	i		1.9		
R. tomentosa	i	i		2.1		
R. sherardii	i					
R. mollis	i	i				
R. rubiginosa	i	i		1.9		
R. micrantha	i					
R. elliptica	i					
R. agrestis	i			2.0		
R. dumalis		i				
R. coriifolia	i					
Agrimonia eupatoria	i	i		1.8		
A. procera	i	i	−4	1.9		
Aremonia agrimonioides	n	e				
Sanguisorba officinalis	i	i				
S. minor	i	n				
Acaena novae-zelandiae	n					
Dryas octopetala	i	i		0.3		+27
Geum rivale	i	i		0.6		
G. urbanum	i	i		1.6		
G. hispidum		i				
Potentilla fruticosa	i	i				
P. palustris	i	i		0.6		
P. anserina	i	i				
P. egedii		i				
P. rupestris	i	i		2.0		
P. multifida		i				
P. nivea		i		0.4		+25
P. chamissonis		i				
P. argentea	i	i		1.1		

Species	Status		Climatic parameters			
	BI	Fe	Wl	R	Wh	Tmax
P. collina		i				
P. norvegica	n	i				
P. intermedia	n	n		1.6		
P. recta	n	n		1.8		
P. thuringiaca		n				
P. crantzii	i	i			−1	+29
P. tabernaemontanii	i	i				
P. erecta	i	i				
P. anglica	i	i	−4			
P. heptaphylla		i				
P. sterilis		i	−1			
P. hyparctica		i				
P. cinerea		i		1.9		
P. reptans		n				
P. sterilis	i					
Sibbaldia procumbens	i	i		0.2		+25
Fragaria vesca	i	i		0.9		
F. viridis		i		1.9		
Alchemilla alpina	i	i		0.5		+26
A. conjuncta	n					
A. glaucescens	i	i		1.7	0	
A. hirsuticaulis		i				
A. plicata		i			−2	
A. monticola	i	i		1.4	−1	
A. propinqua		i				
A. subglobosa		i			−2	
A. sarmatica		i				
A. subcrenata		i			−1	
A. cymatophylla		i				
A. heptagona		i			−4	
A. acutiloba	i	i		1.6		
A. xanthochlora	i	i	−3			
A. gracilescens	i					
A. filicaulis	i	i		0.7	−1	+31
A. vestita	i	i				+31
A. minima*	i					
A. glomerulans	i	i		0.5		+28
A. borealis*		i				

Species	Status		Climatic parameters			
	BI	Fe	Wl	R	Wh	Tmax
A. baltica		i				
A. wichurae	i	i				+27
A. murbeckiana		i			−4	+29
A. glabra	i	i			0	+31
*A. oxyodonta**		i				
A. obtusa		i				
Aphanes arvensis	i	n	−3	1.9		
A. microcarpa	i	n	−3	1.9		
Pyrus pyraster	n			2.0		
P. cordata	n?					
Malus sylvestris	n	n		1.8		
Sorbus aucuparia	i	i		0.8		
Amelanchier spicata		n				
A. grandiflora	n	n				
Cotoneaster integerrimus	i	i			−1	+31
C. cinnabarinus		i			−11	
Crataegus calycina		i	−5	1.8		
C. monogyna	i	i		1.8		
C. laevigata		n	−3	1.9		
Prunus spinosa	i	i		1.9		
P. domestica	n	n				
P. avium	i	i	−4			
P. cerasus	n	n				
P. padus	i	i		0.9	0	+33
P. laurocerasus	n		+4			

FABACEAE

Cytisus scoparius	i	n	−6			
Genista tinctoria	i	i		1.9		
G. anglica	i	i	−1			
G. pilosa	i	i	−3			
G. germanica		i				
Ulex europaeus	i	n	−2	2.0		
U. minor	i	e				

Species	Status		Climatic parameters			
	BI	Fe	Wl	R	Wh	Tmax
U. gallii		i		+5		
Lupinus polyphyllus	n	n				
Robinia pseudacacia	n	e		2.2		
Astragalus danicus	i	i				
A. frigidus		i				+25
ssp. *grigorjewii**		i				
A. penduliflorus		i				
A. alpinus	i	i		0.4		+28
A. norvegicus		i				+26
A. glycyphyllos	i	i		1.7		
A. arenaria		i				
Oxytropis lapponica		i		0.5		+19
O. campestris	i	i				+28
O. halleri	i					
O. deflexa		i				
O. pilosa		i		2.0		
Vicia orobus	i	i	0			
V. pisiformis		i		2.0	−2	
V. cracca	i	i				
V. tenuifolia	e	i		2.0		
V. cassubica		i		1.9		
V. sylvatica	i	i			0?	
V. dumetorum		i			−1	
V. hirsuta	i	i				
V. tetrasperma	i	i		1.6		
V. sepium	i	i		1.1		
V. lutea	i	e				
V. pannonica	i	n				
V. sativa	i	e				
V. villosa	e	n				
V. lathyroides	i	i				
V. bithynica	i	e				
Lathyrus vernus		i		1.5		
L. niger	i	i	−8	1.8		
L. japonicus	i	i				
L. montanus	i	i	−8			

Species	Status		Climatic parameters			
	BI	Fe	Wl	R	Wh	Tmax
L. nissolia	i					
L. hirsutus	n	e				
L. pratensis	i	i				
L. palustris	i	i				
L. tuberosus	n	n		1.9		
L. sylvestris	i	i		1.7		
L. sphaericus		i				
L. heterophylla		n				
L. aphaca	n	e		2.2		
Ononis repens	i	i	−6	1.8		
O. spinosa	i	i	−5	2.0		
O. arvensis		i		1.8		
O. reclinata	i		+6			
Melilotus altissima	n	n				
M. officinalis	n	n				
M. alba	n	n				
Medicago falcata	i	n				
M. lupulina	i	i				
M. minima	i	i		2.1		
M. polymorpha	i	e				
M. arabica	i	e				
Trifolium aureum	n	n		1.7		
T. spadiceum		n			−2	
T. campestre	i	i				
T. dubium	i	i		1.7		
T. micranthum	i	i		2.0		
T. repens	i	n				
T. hybridum	n	n				
T. montanum		i			0	
T. fragiferum	i	i				
T. pratense	i	i				
T. medium	i	i		1.5		
T. arvense	i	i		1.7		
T. striatum	i	i				
T. ornithopodioides	i		+4			
T. strictum	i					
T. occidentale	i		+7			

Species	Status		Climatic parameters			
	BI	Fe	Wl	R	Wh	Tmax
T. bocconei	i		+7			
T. scabrum	i					
T. incarnatum ssp. *incarnatum*	n	e				
ssp. *molinieri*	i		+7			
T. ochroleucon		n				
T. squamosum	i		+4			
T. subterraneum	n		+4			
T. suffocatum	i		+4			
Lotus corniculatus	i	i		0.7		
L. tenuis	i	i				
L. uliginosus	i	i	−4			
L. subbiflorus	i		+5			
L. angustissimus	i	e	+5			
Tetragonolobus maritimus	n	i				
Anthyllis vulneraria	i	i		0.9		
Ornithopus perpusillus	i	i				
O. pinnatus	i					
Coronilla emerus		i		2.0		
C. varia	n	e				
Hippocrepis comosa	i					

OXALIDACEAE

Oxalis acetosella	i	i				
O. corniculata	n	i		1.9		
O. europaea	n	n				

GERANIACEAE

Geranium sanguineum		i		1.9		
G. macrorrhizum		e				
G. pratense	i	i				
G. sylvaticum	i	i			0	+33
G. endresii	n					
G. palustre		i				
G. pyrenaicum	i	n				
G. bohemicum		i		1.8		

Species	Status		Climatic parameters			
	Bl	Fe	Wl	R	Wh	Tmax
G. lanuginosum		i				
G. molle	i	i	−7			
G. pusillum	i	i				
G. columbinum	i	i	−7			
G. rotundifolium	i	e				
G. dissectum	i	i	−4	1.9 as indigenous		
G. lucidum	i	i	−3			
G. robertianum	i	i		1.4		
G. purpureum	i		+5			
G. versicolor	n					
G. nodosum	n					
G. phaeum	n	e				
G. rotundifolium	i	e				
Erodium cicutarium	i	n				
E. moschatum	i	e				
E. maritimum	i					

LINACEAE

Linum bienne	i					
L. usitatissimum	n	e				
*L. perenne**	i					
L. catharticum	i	i		1.6		
L. austriacum		n				
Radiola linoides	i	i		1.9		

EUPHORBIACEAE

Mercurialis annua	n	n				
M. perennis	i	i		1.8 aberrant in N Norway		
Euphorbia peplis	i		+5			
E. lathyris	n	e				
E. villosa	n					
E. corallioides		n				
E. hyberna	i		+6			
E. dulcis	n	e				
E. platyphyllos	i					
E. serrulata	i					

Species	Status		Climatic parameters			
	Bl	Fe	Wl	R	Wh	Tmax
E. helioscopia	i	n				
E. peplus	i	n				
E. exigua	i	n		2.0		
E. portlandica		i				
E. paralias	i					
E. esula	n	n				
E. cyparissias	n	n		1.9		
E. amygdaloides	i	e		2.2		
E. palustris		i				

POLYGALACEAE

Polygala vulgaris	i	i				
P. serpyllifolia	i	i	0			
P. amarella	i	i				
P. calcarea	i			2.3		
P. comosa		i		1.9		

ACERACEAE

Acer platanoides	n	i		1.8		
A. campestre	i	i		2.0		
A. pseudoplatanus	n	n		2.0		

BALSAMINACEAE

Impatiens noli-tangere	i	i		1.6		

AQUIFOLIACEAE

Ilex aquifolium	i	i	0			

CELASTRACEAE

Euonymus europaeus	i	n		2.0		

BUXACEAE

Buxus sempervirens	i	e		2.3		

RHAMNACEAE

Rhamnus cathartica	i	i				

Species	Status		Climatic parameters			
	BI	Fe	Wl	R	Wh	Tmax
Frangula alnus	i	i				
TILIACEAE						
Tilia cordata	i	i				
T. platyphyllos	i?	n				
MALVACEAE						
Malva moschata	i	n				
M. sylvestris	i	n				
M. nicaeansis		n				
M. neglecta	i	n				
M. pusilla	n	n				
M. pauciflora	n					
M. verticillata	n	n				
M. alcea	n	n				
Lavatera arborea	i		+5			
L. cretica	i		+7			
L. thuringiaca		n				
THYMELEACEAE						
Daphne mezereum	i	i			-1	+33
D. laureola	i			2.0		
ELAEAGNACEAE						
Hippophae rhamnoides	i	i				
HYPERICACEAE						
Hypericum androsaemum	i		+4			
H. inodorum	n		+4			
H. hircinum	n					
H. calycinum	n					
H. perforatum	i	i		1.7		
H. maculatum	i	i		1.2		+33
H. undulatum	i		+5			
H. tetrapterum	i	i				
H. humifusum	i	i				

Species	Status		Climatic parameters			
	BI	Fe	Wl	R	Wh	Tmax
H. lineariifolium	i		+5			
H. pulchrum	i	i	−1			
H. hirsutum	i	i				
H. montanum	i	i	−4			
H. elodes	i		+4			
H. canadense	n					
VIOLACEAE						
Viola odorata	i	n	1.9			
V. hirta	i	i	1.9			
V. rupestris	i	i		−1		
V. riviniana	i	i	1.4			
V. reichenbachiana	i	i	1.9			
V. canina	i	i				
V. montana	i	i		−3		
V. lactea	i		+4			
V. persicifolia	i	i	1.8			
V. palustris	i	i	0.5			
V. epipsila		i				
V. lutea	i					
V. tricolor	i	i				
V. curtisii	i					
V. arvensis	i	n	1.4			
V. kitaibeliana	i					
V. biflora		i	0.3			+26
V. alba		i				
V. collina		i	1.7	−3		
V. mirabilis		i	1.6	−1		
V. uliginosa		i				
V. selkirkii		i		−7		
V. pumila		i				
V. elatior		i				
CISTACEAE						
Helianthemum nummularium	i	i				
H. apenninum	i					
H. canum	i	i				
*H. levigatum**	i					

Species	Status		Climatic parameters			
	BI	Fe	Wl	R	Wh	Tmax
H. oelandicum*		i				
Fumana procumbens		i				
Tuberaria guttata	i					
TAMARICACEAE						
Myricaria germanica		i				
Tamarix gallica	n					
FRANKENIACEAE						
Frankenia laevis	i					
ELATINACEAE						
Elatine hexandra	i	i	−3			
E. hydropiper	i	i				
E. triandra		i				
E. alsinastrum		i				
CUCURBITACEAE						
Bryonia cretica	i	e				
LYTHRACEAE						
Lythrum salicaria	i	i		1.6		
L. hyssopifolia	i	i				
L. portula	i	i		1.8		
TRAPACEAE						
Trapa natans		i				
ONAGRACEAE						
Circaea lutetiana	i	i		1.8		
C. intemedia	i	i			−1	
C. alpina	i	i			−1	+32
Ludwigia palustris	i		+5			

Species	Status		Climatic parameters			
	BI	Fe	Wl	R	Wh	Tmax
Epilobium hirsutum	i	i		1.9		
E. parviflorum	i	i		1.9		
E. montanum	i	i				
E. lanceolatum	i	e		2.2		
E. roseum	i	i		1.9		
E. ciliatum	n	n				
E. tetragonum	i	i		1.9		
E. obscurum	i	i		1.9		
E. palustre	i	i		0.8		
E. anagallidifolium	i	i		0.3		+25
E. alsinifolium	i	i		0.8		+27
E. brunnescens	n	n				
E. collinum		i			−2	+29
E. glandulosum		n		1.7		
E. saximontanum		n				
E. lamyi		i				
E. lactiflorum		i		0.6	−7	+27
E. hornemannii		i		0.6	−7	+28
E. davuricum		i		0.7		
Chamaenerion angustifolium	i	i		0.4		

HALORAGACEAE

Myriophyllum verticillatum	i	i				
M. spicatum	i	i				
M. alterniflorum	i	i				
M. exalbescens	i	i				

HIPPURIDACEAE

Hippuris vulgaris	i	i				
H. tetraphylla		i				

CORNACEAE

Cornus sanguinea	i	i		2.0		
C. suecica	i	i				

ARALIACEAE

Hedera helix	i	i	−3			

Species	Status		Climatic parameters			
	BI	Fe	Wl	R	Wh	Tmax
APIACEAE						
Hydrocotyle vulgaris	i	i	–4			
Sanicula europaea	i	i				
Eryngium maritimum	i	i				
E. campestre	i					
Chaerophyllum temulentum	i	n		>2.1		
C. hirsutum				2.3		
Anthriscus sylvestris	i	i				
A. caucalis	i	n				
Scandix pecten-veneris	n	n				
Myrrhis odorata	n	n				
Bunium bulbocastanum	i	e		2.3		
Conopodium majus	i	i	0			
Pimpinella major	i	i		1.8		
P. saxifraga	i	i				
Aegopodium podagraria	n	n		1.6		
Sium latifolium	i	i				
Berula erecta	i	i		1.9		
Crithmum maritimum	i					
Seseli libanotis	i	i				
Oenanthe fistulosa	i	i				
O. pimpinelloides	i		+5			
O. silaifolia	i					
O. lachenalii	i	i				
O. crocata	i					
O. aquatica	i	i		1.9		
O. fluviatilis	i		+4	2.1		
Aethusa cynapium	i	n				
Silaum silaus	i	i		1.9		
Foeniculum vulgare	i?	e				

Species	Status		Climatic parameters			
	BI	Fe	Wl	R	Wh	Tmax
Meum athamanticum	i	n				+28
Conium maculatum	i	n		1.7		
Pleurospermum austriacum		i			−5	
Physospermum cornubiense	i					
Bupleurum rotundifolium	i	e	+4			
B. subovatum	n					
B. baldense	i		+6			
B. tenuissimum	i	i				
B. falcatum	n			2.2	+5	
Trinia glauca	i			2.4		
Apium graveolens	i	e		1.9		
A. nodiflorum	i	e		2.2		
A. repens	i			2.2		
A. inundatum	i		−2			
Petroselinum crispum	n	e				
P. segetum	i					
Cicuta virosa	i	i				
Sison amoenum	i					
Carum carvi	i	i				
C. verticillatum	i					
Cnidium dubium		i				
Selinum carvifolia	i	i		1.9		
Ligusticum scoticum	i	i				
Conioselinum tataricum		i				
Angelica sylvestris	i	i		0.7		
A. archangelica		i		0.5		
Peucedanum officinale	i	i				
P. palustre	i	i				
P. ostruthium	n	n				
P. oreoselinum	n	n		1.9		
Pastinaca sativa	n	n				

Species	Status		Climatic parameters			
	BI	Fe	Wl	R	Wh	Tmax
Heracleum sphondylium	i	i				
H. sibiricum		i				
Tordylium maximum	n					
Laserpitium latifolium		i		2.0		
L. pruthenicum		i		2.0		
Torilis arvensis	i					
T. japonica	i	i		1.8		
T. nodosa	i					
Daucus carota	i	i		1.8		
DIAPENSIACEAE						
Diapensia lapponica	i	i				+24
PYROLACEAE						
Pyrola minor	i	i		0.5		
P. media	i	i				
P. chlorantha		i				
P. rotundifolia	i	i				
*P. norvegica**		i				+27
Orthilia secunda	i	i				
Moneses uniflora	i	i				
Chimaphila umbellata		i		1.8		
Monotropa hypopitys	i	i		1.8		
M. hypophegea	i	i				
ERICACEAE						
Erica mackaiana	i		+6			
E. tetralix	i	i	−5			
E. cinerea	i	i	0			
E. erigena	i		+6			
E. vagans	i		+7			
E. ciliaris	i					
Calluna vulgaris	i	i		0.6		

Species	Status		Climatic parameters			
	BI	Fe	Wl	R	Wh	Tmax
Cassiope tetragona		i				+20
C. hypnoides		i		0.2		+22
Ledum palustre	n	i				
Rhododendron lapponicum		i				+22
Loiseleuria procumbens	i	i		0.3		+25
Phyllodoce coerulea	i	i		0.2		+27
Daboecia cantabrica	i		+6			
Arbutus unedo	i?		+6			
Arctostaphylos uva-ursi	i	i		0.5	0	
A. alpina	i	i		0.5		+25
Andromeda polifolia	i	i		0.6		
Chamaedaphne calyculata		i			−4	
Vaccinium oxycoccus	i	i				
V. microcarpum	i	i			−2	
V. vitis-idaea	i	i		0.3		
V. uliginosum	i	i		0.4		
V. myrtillus	i	i		0.3	+2	
EMPETRACEAE						
Empetrum nigrum	i	i			−3	
E. hermaphroditum	i	i		0.3	0	+29
PRIMULACEAE						
Primula farinosa	i	i			−1	
*P. scotica**	i					
*P. scandinavica**		i				+22
P. stricta		i				+24
P. nutans		i				
P. veris	i	i		1.4		
P. elatior	i	i				
P. vulgaris	i	i	−2			
Androsace septentrionalis		i			−1	
Hottonia palustris	i	i		1.9		

Species	Status		Climatic parameters			
	BI	Fe	Wl	R	Wh	Tmax
Lysimachia nemorum	i	i	−2			
L. nummularia	i	i		1.7		
L. vulgaris	i	i		1.7		
L. thyrsiflora	i	i			+1	
Trientalis europaea	i	i		0.7	+1	
Glaux maritima	i	i				
Anagallis tenella	i		+2			
A. arvensis	n	n				
A. minima	i	i				
Samolus valerandi	i	i		1.9		
PLUMBAGINACEAE						
Armeria maritima	i	i				
A. alliacea	i					
A. scabra		i				
Limonium vulgare	i	i				
L. humile	i	i				
L. bellidifolium	i					
L. auriculae-ursifolium	i					
L. binervosum	i					
*L. recurvum**	i					
*L. transwallianum**	i					
*L. paradoxum**	i					
OLEACEAE						
Fraxinus excelsior	i	i		1.9		
Ligustrum vulgare	i	i	−3	2.0		
GENTIANACEAE						
Cicendia filiformis	i		0	2.0		
Blackstonia perfoliata	i					
Exaculum pusilla	i					
Centaurium pulchella	i	i				
C. tenuiflorum	i					

Species	Status		Climatic parameters			
	BI	Fe	Wl	R	Wh	Tmax
C. erythraea	i	i				
C. littorale	i	i				
C. scilloides	i					
Gentiana pneumonanthe	i	i				
G. verna	i			−1		
G. nivalis	i	i		0.4		+26
G. purpurea		i				+23
Gentianella campestris	i	i			0	
G. germanica	i					
G. amarella	i	i				
*G. anglica**	i					
G. uliginosa	i	i				
G. detonsa		i				
G. aurea		i				
G. tenella		i				+22
*G. baltica**		i				

MENYANTHACEAE

Menyanthes trifoliata	i	i		0.8		

RUBIACEAE

Sherardia arvensis	n	n				
Asperula cynanchica	i					
A. occidentalis	i					
A. taurina	n					
A. tinctoria		i				
Galium odoratum	i	i		1.7		
G. boreale	i	i			−1	
G. album	i	n				
G. verum	i	i				
G. saxatile	i	i	−3			
G. pumilum	i	i				
G. fleurotii	i					
G. sterneri	i	i				
G. normanii		i				
G. palustre	i	i				
G. elongatum	i					

Species	Status		Climatic parameters			
	BI	Fe	Wl	R	Wh	Tmax
G. debile	i		+6			
G. uliginosum	i	i				
G. tricornutum	n	e		2.2		
G. aparine	i	i				
G. spurium	n	n				
G. parisiense	i					
G. trifidum		i				+30
G. triflorum		i			−5	
G. rotundifolium		i				
Cruciata laevipes	i	i		2.1		
Rubia peregrina	i		+5			

POLEMONIACEAE

Species	Status		Climatic parameters			
Polemonium coeruleum	i	i			−5?	
P. acutiflorum		i			−13	
P. boreale		i				

CONVOLVULACEAE

Species	Status		Climatic parameters			
Convolvulus arvensis	i	i		1.7		
Calystegia sepium	n	i		1.8		
C. soldanella	i					
Cuscuta europaea	i	i		1.7		
C. epilinum	n	n				
C. epithymum	i	i		1.9		
C. campestris	n	e				

BORAGINACEAE

Species	Status		Climatic parameters			
Lithospermum officinale	i	i				
L. arvense	i	n		1.5		
L. purpurocoeruleum	i			2.2		
Echium vulgare	i	i				
E. plantagineum	i	i	+7			
Pulmonaria longifolia	i	i				
P. officinale	n					
P. angustifolia		i		1.9		

Species	Status		Climatic parameters			
	BI	Fe	Wl	R	Wh	Tmax
Symphytum officinale	i	n				
S. asperum	n	n				
S. orientale	n					
S. tuberosum		i				
S. ibericum	n					
Anchusa officinale		i				
A. arvensis	i	n				
Mertensia maritima	i	i				
Asperugo procumbens	n	n				
Myosotis scorpioides	i	i				
M. secunda	i					
M. laxa	i	i				
M. sicula	i					
M. alpestris	i	e				
M. sylvatica	i	i			0	
M. arvensis	i	i				
M. discolor	i	i	−5	1.9		
M. ramosissima	i	i	−6	1.8		
M. baltica		i				
M. decumbens		i		0.5		+25
M. sparsiflora		n				
Lappula squarrosa		n				
L. deflexa		i				+30
Cynoglossum officinale	i	i		1.7		
C. germanicum	i					
Omphalodes verna	n	e				

CALLITRICHACEAE

Callitriche stagnalis	i	i				
C. platycarpa	i	i				
C. obtusangula	i					
C. cophocarpa	i	i				
C. palustris	n	i				
C. hamulata	i	i				
C. brutia	i	i	0			

Species	Status		Climatic parameters			
	BI	Fe	Wl	R	Wh	Tmax
C. hermaphroditica	i	i				
C. truncata	i					
LAMIACEAE						
Ajuga pyramidalis	i	i		0.9	0	+32
A. chamaepitys	i					
A. reptans	i	i		1.7		
A. genevensis	e	e		2.1		
Teucrium scorodonia	i	i	−1			
T. scordium	i	i		2.0		
T. chamaedrys	n					
T. botrys	i					
Scutellaria galericulata	i	i		1.3		
S. hastifolia	n	i				
S. minor	i	i	−1			
Marrubium vulgare	i	n		1.9		
Melittis melissophylla	i					
Galeopsis angustifolia	'n	i		1.9		
G. ladanum	n	n				
G. segetum	i	e				
G. tetrahit	i	n				
G. bifida	i	i				
G. speciosa	n	n				
Lamium amplexicaule	i	n				
L. moluccellifolium	i	n				
L. hybridum	i	n				
L. purpureum	i	n				
L. album	i	n				
L. maculatum	n	e				
L. galeobdolon	i	i		1.9		
Leonurus cardiaca	n	n		1.7		
Ballota nigra	i	n				
Stachys annua	i	e				
S. arvensis	i	n				
S. germanica	i	e				

Species	Status		Climatic parameters			
	BI	Fe	Wl	R	Wh	Tmax
S. palustris	i	i		1.4		
S. sylvatica	i	i				
S. officinale	i	n				
Nepeta cataria	n	n		1.8		
Glechoma hederacea	i	i		1.4		
Dracocephalum ruyschiana		i			−3	
Prunella vulgaris	i	i		0.9		
P. grandiflora		i				
P. laciniata	n			2.0		
Acinos arvensis	i	i		1.6		
Clinopodium vulgare	i	i		1.8		
Origanum vulgare	i	i		1.7		
Calamintha nepeta	i					
Thymus pulegioides	i	i				
T. praecox	i	i				
T. serpyllum	i	i				
Lycopus europaeus	i	i		1.8		
fossil				1.4		
Mentha pulegium	i					
M. arvense	i	i				
M. aquatica	i	i		1.8		
M. spicata	n	n				
M. suaveolens	i	n				
M. longifolia		i				
Salvia verticillata	n	e				
S. pratensis	i	n				
S. verbenaca	i					

SOLANACEAE

Atropa belladonna	i	e				
Hyoscyamus niger	i	i		1.4		
Solanum nigrum	i	n				

Species	Status		Climatic parameters			
	BI	Fe	Wl	R	Wh	Tmax
S. dulcamara	i	i		1.7		
S. luteum		n				
Datura stramonium	n	n		1.8		

SCROPHULARIACEAE

Species	Status		Climatic parameters			
Limosella aquatica	i	i				
L. australis	i					
Mimulus guttatus	n	n				
M. luteus	n					
M. moschatus	n	e				
Sibthorpia europaea	i		+5			
Erinus alpinus	n					
Digitalis purpurea	i	i	−1			
Verbascum thapsus	i	i		1.7		
V. phlomoides	n	e				
V. lychnitis	i	e				
V. pulverulentus	i					
V. nigrum	i	i		1.7		
V. blattaria	n	e				
V. virgatum	i	e				
V. densiflorum		i		1.9		
Scrophularia nodosa	i	i		1.7 but wider in N Norway		
S. auriculata	i	e				
S. umbrosa	i	e		2.0	0?	
S. scorodonia	i	e	+6			
S. vernalis	n	e				
Antirrhinum majus	n	e				
Misopates orontium	n	n				
Chaenorrhinum minus	i	n				
Linaria pelisseriana	i		+7			
L. purpurea	n					
L. supina	n	e				

Species	Status		Climatic parameters			
	BI	Fe	Wl	R	Wh	Tmax
L. arenaria	n					
L. repens	n	n				
L. vulgaris	i	i		1.3		
Cymbalaria muralis	n	e				
Kickxia spuria	i	n				
K. elatine	i	e				
Veronica beccabunga	i	i		1.7		
V. anagallis-aquatica	i	i		1.7		
V. catenata	i	i				
V. scutellata	i	i				
V. officinalis	i	i				
V. montana	i	i		2.0		
V. chamaedrys	i	i				
V. spicata	i	i		1.9		
V. fruticans	i	i		0.4		+25
V. repens	n	e				
V. alpina	i	i		0.3		+25
V. serpyllifolia	i	i				+25?
V. humifusa	i	i				
V. peregrina	n	e		2.0		
V. arvensis	i	i		1.5		
V. verna	i	i		1.6		
V. acinifolia	n					
V. praecox	n	e				
V. triphyllos	n	i				
V. hederifolia	i	n		2.0		
V. persica	n	n				
V. polita	i	n				
V. agrestis	i	n				
V. filiformis	n	e				
V. longifolia		i				
V. opaca		n				
Melampyrum cristatum	i	i		1.9		
M. arvense	i	i		1.9	+2	
M. pratense	i	i		0.9		
M. sylvaticum	i	i		0.8	−1	+29
M. nemorosum		i				

Species	Status		Climatic parameters			
	BI	Fe	Wl	R	Wh	Tmax
Euphrasia rostkoviana	i	i		1.7	−1?	
*E. rivularis**	i					
*E. anglica**	i					
*E. vigursii**	i					
E. borealis	i	i				
E. tetraquetra	i					
*E. pseudokerneri**	i					
E. confusa	i					
E. frigida	i	i		0.4		+27
*E. foulaensis**	i					
*E. cambrica**	i					
E. ostenfeldii	i					
*E. màrschallii**	i					
*E. rotundifolia**	i					
*E. campbelliae**	i					
E. scottica	i	i	−1			
*E. heslop-harrisonii**	i					
E. salisburgensis	i	i				
*E. hyperborea**		i				+23
E. curta		i				
*E. bottnica**		i				
E. micrantha		i	−5			
*E. lapponica**		i				
Odontites verna	i	i		1.7		
O. litoralis		i				
O. vulgaris		i				
Parentucellia viscosa	i	e	+4			
Bartsia alpina	i	i		0.3		+28
Pedicularis palustris	i	i		0.4		
P. sylvatica	i	i	−4			
P. hirsuta		i				+17
P. lapponica		i		0.4		+28
P. oederi		i				+23
P. flammea		i				+18
P. sceptrum-carolinum		i			−1	
Rhinanthus groenlandicus		i				
R. minor	i	i		0.7		

Species	Status		Climatic parameters			
	BI	Fe	Wl	R	Wh	Tmax
R. angustifolius	i	n				
Lathraea squamaria	i	i		1.9		
GLOBULARIACEAE						
Globularia vulgaris		i				
OROBANCHACEAE						
Orobanche purpurea	i	i		2.0		
O. rapum-genistae	i	i				
O. alba	i	i				
O. caryophyllacea	i	e				
O. elatior	i					
O. reticulata	i	i				
O. minor	n	n		2.0		
O. loricata	i					
O. hederae	i	e	+5			
LENTIBULARIACEAE						
Pinguicula lusitanica	i		+4			
P. alpina	i	i				+28
P. vulgaris	i	i		0.4	0	
P. grandiflora	i		+6			
P. villosa		i			−10	
Utricularia vulgaris	i	i				
U. australis	i	i		1.9		
U. intermedia	i	i				
U. minor	i	i				
U. ochroleuca	?	i				
PLANTAGINACEAE						
Plantago major	i	i				
P. media	i	i				
P. lanceolata	i	i				
P. maritima	i	i				
P. coronopus	i	i				
P. arenaria	n	e				

Species	Status		Climatic parameters			
	BI	Fe	Wl	R	Wh	Tmax
P. tenuiflora		i				
Littorella uniflora	i	i				
CAPRIFOLIACEAE						
Sambucus ebulus	n	e				
S. nigra	i	n				
S. racemosa	n	n				
Viburnum opulus	i	i		1.6		
V. lantana	i	e		2.1		
Lonicera periclymenum	i	i	–2			
L. caprifolium	n	e				
L. xylosteum	i	i		1.7		
L. caerulea		i				
Linnaea borealis		i		0.7	–1	+32
ADOXACEAE						
Adoxa moschatellina	i	i				
VALERIANACEAE						
Valerianella locusta	i	i				
V. carinata	i	e				
V. rimosa	i	e				
V. eriocarpa	n	e				
V. dentata	i	n				
Valeriana officinalis	i	n				
V. sambucifolia	i	i		0.7		
V. pyrenaica	n					
V. dioica	i	i		2.0		
V. salina		i				
Centranthus ruber	n					
DIPSACACEAE						
Dipsacus fullonum	i	e				
D. sativus	n					
D. pilosus	i					

Species	Status		Climatic parameters			
	BI	Fe	Wl	R	Wh	Tmax
D. strigosus	n	n		2.0		
Succisa pratensis	i	i				
Knautia arvensis	i	n				
Scabiosa columbaria	i	i				
S. canescens	i					
CAMPANULACEAE						
Campanula latifolia	i	i			−1?	
C. trachelium	i	i		1.9		
C. rapunculoides	n	n		1.8		
C. persicifolia	n	i				
C. glomerata	i	n				
C. rotundifolia	i	i		0.5		
C. patula	i	n				
C. rapunculus	n	e				
C. medium	n					
C. barbata		i				
C. cervicaria		i		1.8	−1	
C. uniflora		i				+17
Phyteuma spicata	i	i?				
P. orbiculare	i	e				
Wahlenbergia hederacea	i		+4			
Jasione montana	i	i		1.8		
Lobelia urens	i	e	+5			
L. dortmanna	i	i				
ASTERACEAE						
Eupatorium cannabinum	i	i		1.9		
Solidago virgaurea	i	i		0.4		
S. canadensis	e	n				
Bellis perennis	i	i				
Aster tripolium	i	i				
A. linosyris	i	i		2.0		

Species	Status		Climatic parameters			
	BI	Fe	Wl	R	Wh	Tmax
A. sibiricus		i			−10	
Erigeron acer	i	i				+26
E. borealis	i	i		0.6		
E. karwinskianus	n		+4			
E. politus		i				
E. uniflorus		i		0.2		+19
E. eriocephalus		i				
E. humilis		i				+18
Filago vulgaris	i	i				
F. lutescens	i	i		2.1		
F. pyramidata	i					
F. gallica	n					
F. minima	i	i				
F. arvensis		i				
Omalotheca sylvatica	i	i		1.2		
O. norvegica	i	i		0.4	−5	+27
O. supina	i	i		0.2		+26
Filaginella uliginosa	i	i				
Gnaphalium luteoalbum	i	e		2.0		
Anaphalis margaritacea	n	e				
Antennaria dioica	i	i		0.4		
*A. nordhageniana**		i				
*A. alpina**		i		0.1		+25
A. porsildii		i				+18
A. villifera		i				+23
Inula helenium	n	e				
I. salicina	i	i				
I. conyza	i			2.1		
I. crithmoides	i					
I. britannica	n	n				
I. erucifolia		i				
Pulicaria dysenterica	i	e		2.0		
P. vulgaris	i	e		2.0		
Bidens cernua	i	i		1.7		

Species	Status		Climatic parameters			
	BI	Fe	Wl	R	Wh	Tmax
B. connata	n					
B. tripartita	i	i		1.6		
B. radiata		i				
Ambrosia artemisiifolia	n	e				
A. coronopifolia	n	e				
A. trifida	n	e				
Xanthium spinosum	n	e				
X. ambrosioides	n	e				
X. strumarium	n	e		>2.1		
Galinsoga parviflora	n	n		>2.1		
G. ciliata	n	n				
Anthemis tinctoria	n	i				
A. arvensis	i	n				
A. cotula	i	n		>2.1		
A. nobilis	n	e	+4			
Achillea ptarmica	i	i				
A. millefolium	i	i				
A. nobilis		n		2.1		
Matricaria maritima	i	i				
M. inodora	i	n				
M. matricarioides	n	n				
M. recutita	i	n				
Tanacetum vulgare	i	n				
T. parthenium	n	e				
Chrysanthemum segetum	n	n				
Leucanthemum vulgare	i	i				
Cotula coronopifolia	n	n				
Artemisia vulgaris	i	i				
A. verlotiorum	n					
A. stellerana	n					
A. norvegica	i	i				
A. absinthium	n	n				
A. maritima	i	i				
A. campestris	i	i		1.8		

Species	Status		Climatic parameters			
	BI	Fe	Wl	R	Wh	Tmax
A. rupestris		i				
*A. oelandica**		i				
*A. bottnica**		i				
Tussilago farfara	i	i				
Petasites hybridus	i	n				
P. albus	n	n				
P. japonicus	n	n				
P. fragrans	n					
P. frigidus		i	0.3	−9		+29
P. spurius		i				
Homogyne alpina	n					
Arnica alpina		i				+26
A. montana		i				
Senecio jacobaea	i	i		1.7		
S. aquaticus	i	i	−2			
S. erucifolius	i	n		2.0		
S. squalidus	n	e				
*S. cambrensis**	i					
S. viscosus	n	n		1.7		
S. vulgaris	i	i				
S. paludosus	n	i		1.9		
S. fluviatilis	n					
S. congestus	i	n				
S. integrifolius	i	i				
S. sylvaticus	i	i		2.7		
S. vernalis		n				
S. nemorensis		n				
Carlina vulgaris	i	i		1.9		
Arctium lappa	i	n		1.8		
A. minus	i	n				
A. nemorosum	i	i				
A. pubescens	i					
A. tomentosum		n				
Saussurea alpina	i	i	0.2			+28
Carduus tenuiflorus	i	e				

Species	Status		Climatic parameters			
	BI	Fe	Wl	R	Wh	Tmax
C. pycnocephalus	n					
C. nutans	i	n				
C. acanthoides	i	n				
C. crispus		n				
Cirsium eriophorum	i					
C. vulgare	i	i	1.6			
C. palustre	i	i	1.3			
C. arvense	i	i	1.2			
C. oleraceum	n	i				
C. acaule	i	i	1.9			
C. helenioides	i	i	0.5	−1		+31
C. dissectum	i	e				
C. tuberosum	i					
Onopordon acanthium	n	n				
Serratula tinctoria	i	i	1.9			
Centaurea scabiosa	i	i	1.5			
C. cyanus	i	n				
C. montana	n	e				
C. paniculata	n					
C. jacea	n	i				
C. nigra	i	i				
C. aspera	n	e				
C. calcitrapa	n	e				
C. solstitialis	n	e				
C. phrygia		i				
C. pseudophrygia		i				
Cichorium intybus	n	n	1.6			
Arnoseris minima	i	i				
Hypochoeris radicata	i	i	1.8			
H. glabra	i	n				
H. maculata	i	i			+1	
Leontodon autumnalis	i	i	0.5			
L. hispidus	i	i				
L. taraxacoides		i				
Scorzonera humilis	n	i				

Species	Status		Climatic parameters			
	BI	Fe	Wl	R	Wh	Tmax
Picris echioides	n	e				
P. hieracioides	i	i				
Tragopogon pratensis	n	n				
Sonchus palustris	i	i		2.0		
S. arvensis	i	i				
S. oleraceus	n	n		1.5		
S. asper	n	n				
S. uliginosus		n				
Lactuca serriola	n	e				
L. virosa		n				
L. quercina		i				
L. sibirica		i			−8	
L. tatarica		n				
Mycelis muralis	i	i		1.8 but not in N Norway		
Cicerbita alpina	i	i		0.5	−4	+28
C. macrophylla		n				
Taraxacum dovrense*		i				
T. croceum		i	i		0.3	
Lapsana communis	i	i				
Crepis foetida	i					
C. biennis	n	n		1.8		
C. capillaris	i	i		1.8		
C. tectorum	n	i			0	
C. mollis	i					
C. paludosa	i	i			0	
C. praemorsa		i		1.9	−2	
C. multicaulis		i				

MONOCOTYLEDONAE

BUTOMACEAE

Butomus umbellatus	i	i				

Species	Status		Climatic parameters			
	Bl	Fe	Wl	R	Wh	Tmax
ALISMATACEAE						
Sagittaria sagittifolia	i	i				
S. natans		i			−7	
Baldellia ranunculoides	i	i				
Luronium natans	i	i				
Alisma plantago-aquatica	i	i				
A. lanceolata	i	i		2.0		
A. graminea	n					
A. wahlenbergii*		i				
HYDROCHARITACEAE						
Hydrocharis morsus-ranae	i	i				
Stratiotes aloides	i	i				
Elodea canadensis	n	n				
Hydrilla verticillata	i					
SCHEUCHZERIACEAE						
Scheuchzeria palustris	i	i			−1	
JUNCAGINACEAE						
Triglochin maritima	i	i				
T. palustris	i	i				
POTAMOGETONACEAE						
Potamogeton natans	i	i				
P. polygonifolius	i	i	−5			
P. coloratus	i	i				
P. nodosus	i					
P. lucens	i	i				
P. gramineus	i	i				
P. alpinus	i	i				
P. praelongus	i	i				
P. perfoliatus	i	i				
P. epihydrus		i				

Species	Status		Climatic parameters			
	BI	Fe	Wl	R	Wh	Tmax
P. friesii	i	i				
P. rutilus	i	i				
P. pusillus	i	i				
P. obtusifolius	i	i				
P. berchtoldii	i	i				
P. trichoides	i	i				
P. compressus	i	i				
P. acutifolius	i	i				
P. crispus	i	i		1.9		
P. filiformis	i	i				
P. pectinatus	i	i				
P. panormitanus		i				
P. vaginatus		i				
Groenlandia densa	i					

RUPPIACEAE

Ruppia cirrhosa	i	i				
R. maritima	i	i				

ZOSTERACEAE

Zostera marina	i	i				
Z. noltii	i	i				
Z. angustifolia	i					

ZANNICHELLIACEAE

Zannichellia major		i				
Z. palustris		i				

NAJADACEAE

Najas flexilis	i	i				
N. marina	i	i				
N. tenuissima		i				

LILIACEAE

Tofieldia pusilla	i	i	0.4	-7?	+26	
T. calyculata		i		-3		

Species	Status		Climatic parameters			
	BI	Fe	Wl	R	Wh	Tmax
Narthecium ossifragum	i	i				
Veratrum album		i				
Anthericum ramosum		i				
A. liliago		i				
Lloydia serotina	i					
Gagea lutea	i	i			−1?	
G. bohemica	i					
G. pratensis		n				
G. minima		i			−1	
G. arvensis		n				
G. spathacea		i				
Tulipa sylvestris	n	n				
Fritillaria meleagris	i	n				
Ornithogalum umbellatum	n	e				
O. nutans	n	e		2.1		
O. pyrenaicum	i					
Scilla verna	i	i				
Hyacinthoides non-scripta	i	e				
H. hispanica	n	e				
Allium schoenoprasum	i	i				
A. roseum	n					
A. ursinum	i	i				
A. triquetrum	n		+5			
A. paradoxum	n					
A. oleraceum	n	i				
A. ampeloprasum	n		+6−7			
A. scorodoprasum	n	i				
A. vineale	n	i		1.9		
A. carinatum	n	i			0?	
A. senescens		i				
A. strictum		i				
Colchicum autumnale	i	e				
Convallaria majalis	i	i		1.6		

Species	Status		Climatic parameters			
	BI	Fe	Wl	R	Wh	Tmax
Maianthemum bifolium	i	i			0	+33
Asparagus officinalis	i	n				
Polygonatum verticillatum	i	i		0.8		+32
P. odoratum	i	i		1.6		
P. multiflorum	i	i		2.0		
Paris quadrifolia	i	i		1.6		
Ruscus aculeatus	i					

AMARYLLIDACEAE

Leucojum vernum	n	e				
L. aestivum	i					
Narcissus pseudonarcissus	a	e				
N. poeticus	n	e				

IRIDACEAE

Sisyrinchium bermudianum	i					
S. montanum	n	e				
Iris spuria	n	i				
I. foetidissima	i					
I. pseudacorus	i	i		1.6		
I. sibirica		e		1.9		
Romulea columnae	i		+7			
Gladiolus illyricus	i		+6			

DIOSCOREACEAE

Tamus communis	i					

JUNCACEAE

Juncus maritimus	i	i				
J. acutus	i					
J. filiformis	i	i		0.6		
J. balticus	i	i				
J. arcticus		i				+22
J. inflexus	i	i		2.0		

Species	Status		Climatic parameters			
	BI	Fe	Wl	R	Wh	Tmax
J. effusus	i	i		1.8		
J. conglomeratus	i	i				
J. subulatus	n					
J. trifidus	i	i		0.3		+25
J. squarrosus	i	i	−6			
J. compressus	i	i				
J. gerardii	i	i				
J. tenuis	n	n		1.9		
J. bufonius	i	i				
J. ambiguus	i	i				
J. foliosus	i					
J. planifolius	n					
J. capitatus	i	i				
J. subnodulosus	i		−3			
J. pygmaeus	i					
J. bulbosus	i	i				
J. acutiflorus	i	i	−1			
J. articulatus	i	i				
J. alpino-articulatus	i	i			−1	
J. biglumis	i	i		0.3		+22
J. triglumis	i	i		0.5		+25
J. castaneus	i	i				
J. anceps		i				
J. stygius		i			−4	
Luzula campestris	i	i				
L. multiflora	i	i				
L. congesta	i	i	−1			
L. frigida		i				
L. pallescens	i	i				
L. arcuata	i	i				+19
L. spicata	i	i				+24
L. sylvatica	i	i	−1			
L. luzuloides	n	n				
L. pilosa	i	i				
L. forsteri	i			2.3		
L. parviflora		i				
L. wahlenbergii		i				
L. arctica		i				+18
L. sudetica		i			−3	

Species	Status		Climatic parameters			
	BI	Fe	Wl	R	Wh	Tmax
ERIOCAULACEAE						
Eriocaulon aquaticum	i					
POACEAE						
Festuca pratensis	i	i				
F. arundinacea	i	i				
F. gigantea	i	i		1.8		
F. altissima	i	i		1.8		
F. heterophylla	i	n				
F. nigrescens	i					
F. rubra	i	i		0.4		
F. juncifolia	i					
F. tenuifolia		i				
F. ovina	i	i				
F. amoricana	i					
F. guestphalica	i					
F. lemannii	i					
F. longifolia	i					
F. vivipara	i	i		0.1		+26
F. trachyphylla		n				
F. polesica		i				
Lolium perenne	i	n				
L. multiflorum	n	n				
L. temulentum	n	n				
L. remotum		n				
Vulpia fasciculata	i		+4–5			
V. bromoides	i	i				
V. myuros	i	e				
V. ciliata	i					
V. unilateralis	i					
Desmazeria rigida	i	e	+4			
D. marina	i					
Poa annua	i	i				
P. infirma	i		+7			
P. supina		i				
P. bulbosa	i	i				

Species	Status		Climatic parameters			
	BI	Fe	Wl	R	Wh	Tmax
P. alpina	i	i		0.1		
P. flexuosa	i	i				+19
P. nemoralis	i	i				
P. glauca	i	i		0.4		
P. compressa	i	i				
P. pratensis	i	i				
P. angustifolia	i	i				
P. alpigena		i		0.3		
P. subcoerulea	i	i				
P. palustris	i	i				
P. trivialis	i	i				
P. remota		i			−1	
P. chaixii	n	n		1.9		
*P. arctica**		i				+19
*P. stricta**		i				
Arctophila fulva		i				
Puccinellia maritima	i	i				
P. distans	i	i				
P. fasciculata	i					
P. rupestris	i	n				
P. capillaris	i	i				
P. phryganodes		i				
Phippsia algida		i				+20
P. concinna		i				
Arctagrostis latifolia		i				
Dactylis glomerata	i	i				
Cynosurus cristatus	i	i				
Catabrosa aquatica	i	i				
Cinna latifolia		i			−4	+31
Apera spica-venti	n	e				
A. interrupta	n	e				
Mibora minima	i					
Briza media	i	i		1.5		
B. minor	i	e	+6			
B. maxima	n	e				

Species	Status		Climatic parameters			
	BI	Fe	Wl	R	Wh	Tmax
Sesleria albicans	i					
S. coerulea		i				
Melica uniflora	i	i	−3	1.9		
M. nutans	i	i		0.8	0	
Glyceria fluitans	i	i		1.9		
G. plicata	i	i		1.9		
G. declinata	i	i	−2			
G. maxima	i	i				
G. lithuanica		i			−4	
G. striata		n				
Bromus sterilis	i	n		1.8		
B. madritensis	n	e				
B. diandrus	i	e				
B. rigidus	i					
B. tectorum	n	n				
B. inermis	n	n				
B. ramosus	i	i	−3			
B. benekenii	i	i		1.8		
B. hordaceus	i	i				
B. lepidus	n	n				
*B. interruptus**	i					
B. racemosus	i	e				
B. commutatus	i	e				
B. secalinus	i	n				
*B. pseudosecalinus**	n?					
B. arvensis	n	n				
B. erectus	n	n		1.9		
Brachypodium sylvaticum	i	i		1.8		
B. pinnatum	i	i		2.0		
Leymus arenarius	i	i				
Elymus caninus	i	i				
E. repens	i	i				
E. pycnanthus	i					
E. farctus	i	i				
E. mutabilis		i				
E. fibrosus		i				
E. borealis		i				

Species	Status		Climatic parameters			
	BI	Fe	Wl	R	Wh	Tmax
Hordeum secalinum	i	i				
H. murinum	i	n				
H. jubatum		n				
Hordelymus europaeus	i	i		2.0		
Helictotrichon pratense	i	i				
H. pubescens	i	i				
Arrhenatherum elatius	i	i				
Koeleria vallesiaca	i					
K. macrantha	i					
K. glauca	i	i				
K. grande		i				
Trisetum flavescens	i	n				
T. spicatum		i		0.1		+24
T. subalpestre		i				
Lagurus ovatus	n	e	+7			
Vahlodea atropurpurea		i				+26
Deschampsia cespitosa	i	i		0.5		
*D. bottnica**		i				
D. alpina		i				+23
D. flexuosa	i	i				
D. setacea	i	i	−2			
Aira praecox	i	i	−3			
A. caryophyllea	i	i	−3			
Hierochloe odorata	i	i			0	
H. australis		i				
H. hirta		i				
H. alpina		i				
Anthoxanthum odoratum	i	i		0.3		
Holcus lanatus	i	i		1.6		
H. mollis	i	i		1.8		
Corynephorus canescens	i	i				
Agrostis canina	i	i				
A. vinealis	i	i				

Species	Status		Climatic parameters			
	BI	Fe	Wl	R	Wh	Tmax
A. curtisii (setacea)	i		+5	2.2		
A. capillaris	i	i				
A. gigantea	i	i				
A. stolonifera	i	i				
A. mertensii		i		0.4		
A. clavata		i				
Gastridium ventricosum	i					
Polypogon monspeliensis	i	e				
Ammophila arenaria	i	i				
Calamagrostis epigejos	i	i				
C. canescens	i	i				
C. stricta (neglecta)	i	i				
C. scotica*	i					
C. arundinacea		i		1.6	−1	
C. varia		i				
C. chalybaea*		i				
C. lapponica		i			−9	
C. purpurea		i		0.5		
Phleum pratense	i	i				
P. alpinum	i	i		0.4	−4	+28
P. phleoides	i	i				
P. arenarium	i	i				
Alopecurus pratensis	i	i				
A. myosuroides	n	n				
A. geniculatus	i	i				
A. aequalis	i	i				
A. bulbosus	i					
A. alpinus	i					
A. arundinaceus		i				
Parapholis strigosa	i	i				
P. incurva	i					
Phalaris arundinacea	i	i				
Milium effusum	i	i				
M. vernale	i		+7			
Stipa joannis		i				

Species	Status		Climatic parameters			
	BI	Fe	Wl	R	Wh	Tmax
Phragmites australis	i	i				
Danthonia decumbens	i	i	–6			
Molinia coerulea	i	i	–11	0.8		
Nardus stricta	i	i		0.4		+33
Cynodon dactylon	i					
Spartina maritima	i					
*S. townsendii**	i					
*S. anglica**	i					
S. alterniflora	n					
Leersia oryzoides	i					
Digitaria sanguinalis	n	e				
Setaria viridis	n	n	1.8			
ARACEAE						
Calla palustris	e	i				
Acorus calamus	n	n				
Arum maculatum	i		<2.1			
LEMNACEAE						
Wolffia arrhiza	i					
Lemna polyrrhiza	i	i				
L. trisulca	i	i				
L. minor	i	i				
L. gibba	i	i				
SPARGANIACEAE						
Sparganium erectum	i	i	2.0			
S. emersum	i	i				
S. angustifolium	i	i				
S. minimum	i	i				
S. hyperboreum		i				
S. gramineum		i				
S. glomeratum		i				

Species	Status		Climatic parameters			
	BI	Fe	Wl	R	Wh	Tmax
TYPHACEAE						
Typha angustifolia	i	i		1.8		
T. latifolia	i	i		1.7		
CYPERACEAE						
Scirpus sylvaticus	i	i		1.6		
S. radicans		i				
S. maritimus	i	i				
S. triquetrus	i					
S. pungens	i					
S. lacustris	i	i				
S. tabernaemontani	i	i				
S. setaceus	i	i	−3			
S. cernuus	i		+4−5			
S. fluitans	i	i	−1			
S. cespitosus	i	i		0.6		
S. germanicus	i	i				
S. vulgaris	i					
S. hudsonianus	i	i		0.9	0	
S. pumilus		i				
Blysmus compressus	i	i				
B. rufus	i	i				
Eriophorum angustifolium	i	i		0.3		
E. latifolium	i	i		1.3		
E. gracile	i	i				
E. vaginatum	i	i		0.4		+33
E. scheuchzeri		i		0.3	−8	+28
E. russeolum		i			−10	
E. medium		i				
E. brachyantherum		i			−8	
Eleocharis parvula	i	i				
E. acicularis	i	i				
E. quinqueflora	i	i				
E. multicaulis	i	i	−3			
E. palustris	i	i				
E. mamillata	i	i				
E. uniglumis	i	i				

Species	Status		Climatic parameters			
	BI	Fe	Wl	R	Wh	Tmax
Cyperus longus	i	i		2.0		
C. fuscus	i					
Cladium mariscus	i	i		1.9		
Schoenus nigricans	i	i				
S. ferrugineus	i	i				
Rhynchospora alba	i	i		1.5		
R. fusca	i	i		1.6		
Kobresia myosuroides		i		0.4		+24
K. simpliciuscula	i	i				+25
Carex paniculata	i	i				
C. appropinquata	i	i				
C. diandra	i	i				
C. vulpina	i	n				
C. otrubae	i	i				
C. spicata	i	i		1.9		
C. divulsa	i	i	−4	2.0		
C. muricata	i	i				
C. arenaria	i	i				
C. disticha	i	i				
C. divisa	i					
C. chordorrhiza	i	i		0.9	−1	+33
C. maritima	i	i				
C. remota	i	i				
C. ovalis	i	i				
C. macloviana		i				
C. echinata	i	i				
C. dioica	i	i		0.6		
C. parallella		i				+24
C. scirpoidea		i				
C. elongata	i	i				
C. curta	i	i		0.7		
C. lachenalii	i	i		0.2		+26
C. brunnescens		i		0.5	−3	+32
C. lapponica		i			−11	
C. loliacea		i			−3	+33
C. tenuiflora		i			−7	+29
C. disperma		i			−4	+33

Species	Status		Climatic parameters			
	BI	Fe	Wl	R	Wh	Tmax
C. hirta	i	i		1.6		
C. lasiocarpa	i	i		0.9		+33
C. acutiformis	i	i		1.8		
C. riparia	i	i		1.9		
C. pseudocyperus	i	i		1.9		
fossil				1.7		
C. rostrata	i	i		0.6		
C. rhynchophysa	i	i				
C. vesicaria	i	i				
C. saxatilis	i	i		0.3		+23
C. rotundata		i		0.6	−10	+27
C. stenolepis	i	i				
C. pendula	i		−1			
C. sylvatica	i	i		1.8		
C. capillaris	i	i		0.5		
C. strigosa	i		0			
C. flacca	i	i	−5			
C. panicea	i	i		0.8		
C. vaginata	i	i		0.4	−2	+32
C. livida		i			−5	
C. laevigata	i					
C. binervis	i	i	+1	1.9		
C. distans	i	i		2.0		
C. punctata	i	i	−3	2.0		
C. extensa	i	i				
C. hostiana	i	i	−6			
C. flava	i	i		0.7		
C. bergrothii		i				
C. demissa	i	i				
C. serotina	i	i				
C. pulchella		i		0.9		
C. pallescens	i	i		1.5		
C. digitata	i	i			−1	
C. ornithopoda	i	i			−3?	
C. pediformis		i				
C. humilis	i					
C. glacialis		i				+27
C. caryophyllea	i	i		1.8		
C. ericetorum	i	i				

Species	Status		Climatic parameters			
	BI	Fe	Wl	R	Wh	Tmax
C. globularis		i			−3	
C. tomentosa	i	i				
C. montana	i	i		1.9		
C. pilulifera	i	i				
C. misandra		i				+20
C. atrofusca	i	i		0.5		+23
C. limosa	i	i		0.8		
C. magellanica	i	i		0.7	−3?	+29
C. rariflora	i	i		0.6	−9?	+26
C. laxa		i			−10	
C. atrata	i	i				+27
C. buxbaumii	i	i				+28
C. adelostoma		i			−9	
C. hartmanii		i		1.9		
C. norvegica	i	i				+27
C. media		i				+28
C. holostoma		i				
C. stylosa		i				
C. bicolor		i				+22
C. rufina		i				+20
C. paleacea		i				
C. vacillans*		i				
C. halophila*		i				
C. recta (salina)	i	i				
C. subspathacea		i				
C. aquatilis	i	i				
C. bigelowii	i	i		0.1		+27
C. cespitosa		i				
C. nigra	i	i		0.7		
C. juncella		i		0.8	−4	
C. acuta	i	i				
C. microglochin	i	i		0.6		+25
C. pauciflora	i	i			−1	
C. pulicaris	i	i	−4			
C. rupestris	i	i		0.2		+24
C. nardina		i				
C. capitata		i			−9	
C. arctogena		i				+21

Species	Status		Climatic parameters			
	BI	Fe	Wl	R	Wh	Tmax
ORCHIDACEAE						
Cypripedium calceolus	i	i			−1?	
Epipactis palustris	i	i		1.8		
E. helleborine	i	i		1.8		
E. purpurata	i					
E. leptochila	i					
E. dunensis*	i					
E. phyllanthes	i	i				
E. atrorubens	i	i				
Cephalanthera damasonium	i	i		2.1		
C. longifolia	i	i		1.9		
C. rubra	i	i		1.9		
Epipogum aphyllum	i	i				
Neottia nidus-avis	i	i		1.8		
Listera ovata	i	i				
L. cordata	i	i		0.9	0	+32
Spiranthes spiralis	i			2.1		
S. aestivalis	i					
S. romanzoffiana	i					
Goodyera repens	i	i				+33
Herminium monorchis	i	i		1.9		
Platanthera bifolia	i	i		1.3		
P. chlorantha	i	i	−6	1.8		
P. obtusata		i				
Chamorchis alpina		i				+21
Gymnadenia conopsea	i	i		0.7		
G. odoratissima		i				
Pseudorchis albida	i	i				
P. straminea		i		0.5		+22
Nigritella nigra		i				+25
Coeloglossum viride	i	i		0.5		

Species	Status		Climatic parameters			
	BI	Fe	Wl	R	Wh	Tmax
Dactylorhiza sambucina		i				
D. incarnata	i	i				
D. pseudocordigera		i				
D. majalis	i	i				
D. purpurella	i	i	0			
D. traunsteinerioides	i					
D. traunsteineri		i				
D. praetermissa	i	i				
D. cruenta		i				+29
D. maculata	i	i				+33
D. fuchsii	i	i				
Neotinea maculata	i		+6			
Orchis morio	i	i	−4	2.0		
O. ustulata	i	i		2.0		
O. simia	i					
O. militaris	i	i		1.9		
O. spitzelii		i				
O. purpurea	i			2.2		
O. mascula	i	i	−5			
O. laxiflora	i		+7			
Aceras anthropophorum	i					
Himantoglossum hircinum	i					
Anacamptis pyramidalis	i	i		2.0		
Ophrys apifera	i			2.2		
O. fuciflora	i					
O. sphegodes	i					
O. insectifera	i	i		2.2		
Corallorhiza trifida	i	i			−1	
Calypso bulbosa		i			−8	
Liparis loesellii	i	i				
Microstylis monophyllos		i				
Hammarbya paludosa	i	i				

Arctic species of vascular plants

Pleuropogon sabinei
Dupontia fisheri
D. psilosantha
Deschampsia brevifolia
Festuca hyperborea
F. baffinensis
Poa abbreviata
P. alpigena
P. colpodea
Puccinellia angustata
Colpodium vahlianum
Arctagrostis latifolia
Eriophorum triste
Carex ursina
C. amblyrhyncha
C. subspathacea
C. stans

Minuartia rossii
Stellaria ciliatisepala
Cerastium regelii
Eutrema edvardsii
Braya purpurascens
Draba macrocarpa
D. oblongata
D. subcapitata
Saxifraga flagellaris
S. hyperborea
Potentilla pulchella
P. rubricaulis
P. hyparctica
Epilobium arcticum
Cassiope tetragona
Pedicularis lanata coll.

Endemic species of vascular plants, bryophytes and lichens

A Endemics of Fennoscandia and adjacent areas

I Alpine-northern boreal endemics

Vascular plants

Alnus incana ssp. *kolaensis*	*A. alpinus*ssp. *arcticus*
*Rumex acetosa*ssp. *serpentinicola*	*Oxytropis deflexa*ssp. *norvegica*
*Silene uralensis*ssp. *apetala*	*Viola rupestris*ssp. *relicta*
*Thalictrum simplex*ssp. *boreale*	*Pyrola norvegica*
Papaver radicatum several sspp.	*Primula scandinavica*
P. lapponicum several sspp.	*Thymus serpyllum*ssp. *tanaensis*
P. laestadianum	*Euphrasia hyperborea*
Draba dovrensis	*Arnica alpina*ssp. *alpina* ?
D. cacuminum	*Antennaria alpina*
Saxifraga opdalensis	*A. nordhageniana*
Alchemilla borealis	*Taraxacum dovrense*
A. kolaensis ?	*Poa lindbergii* (= *P. stricta*)
A. transpolaris	*P. arctica* several sspp.
Astragalus frigidus ssp. *grigorjewii*	*Elymus alaskanus*ssp. *subalpina*

In addition *Nigritella nigra* is represented by an apomictic race with 64 chromosomes, whereas the *Nigritella* species of the Alps normally are fertile with 40 or 80 chromosomes (Teppner & Klein 1985).

Of the alpine-northern boreal endemics *Draba cacuminum* is a polyploid which probably has arisen more than once in Scandinavia (Brochmann 1992; Brochmann *et al.* 1992). *Primula scandinavica* is possibly an allopolyploid from *P. farinosa* and *P. scotica*. *Saxifraga opdalensis* has arisen from hybridisation between *S. cernua* and *S. rivularis* and reproduces by bulbils. The *Alchemilla* species, *Antennaria alpina*, and the *Poa* taxa are all apomicts.

Liverworts

Marsupella andraeaoides
M. brevissima
Plagiochila norvegica
Scapania sphaerifera
Jungermannia jenseniana

Mosses

Brachythecium ryanii
B. curvatum
Cynodontium suecicum
Fontinalis bryhnii
Gymnostomum boreale
Orthothecium lapponicum (also Svalbard)
Pseudocalliergon angustifolium
Schistidium bryhnii
Sphagnum troendelagicum
Tetraplodon blyttii (also Svalbard, Jan Mayen)

II Halophilous endemics

Vascular plants

   (a) Baltic endemics

Anthyllis vulneraria ssp. *maritima*
Lotus corniculatus var. *maritimus*
Euphrasia bottnica
Odontites litoralis ssp. *fennica*
Rhinanthus serotinus ssp. *halophilus*
   ssp. *arenarius*
Artemisia campestris ssp. *bottnica*
A. maritima ssp. *humifusa*
Alisma wahlenbergii
Deschampsia bottnica
Hierochloe odorata ssp. *baltica*

In addition there are several races along the shores of the Baltic which may deserve taxonomic recognition (Ericsson & Wallentius 1979; Jonsell 1988).

(b) Baltic–West Scandinavian endemics

Polygonum oxyspermum ssp. *oxyspermum*
Atriplex prostrata ssp. *calotheca*
Salicornia dolichostachya ssp. *strictissima*
Cakile maritima ssp. *baltica*
Gentianella baltica
Carex vacillans

(c) North coast endemics

Atriplex lapponica
Salicornia pojarkovae
Carex halophila (also in the Baltic)

(d) South Swedish and Danish sand-dune endemics

Dianthus arenarius ssp. *arenarius*
Euphrasia dunensis
E. arctica ssp. *minor*

III Öland–Gotland (Baltic islands) endemics

Vascular plants

Pulsatilla vulgaris ssp. *gotlandica*
Helianthemum oelandicum ssp. *oelandicum*
H. canum ssp. *canescens*
Euphrasia salisburgensis ssp. *schoenicola*
Rhinanthus rumelicus ssp. *oesiliensis*
Galium oelandicum
Artemisia oelandica
A. maritima ssp. *oelandica*
Saussurea alpina ssp. *esthonica*
Crepis tectorum ssp. *pumila*
Festuca oelandica

Of these *Rhinanthus rumelicus* ssp. *oesiliensis* has its nearest relatives in southeast Europe, while *Artemisia oelandica* has its nearest relatives in the Caucasus and Kazakhstan (Wendelberger 1959).

Bryophytes

Plectocolea gothica
Riccia gothica
R. oelandica
Tortella rigens

IV Other lowland endemics

Vascular plants

Corydalis gotlandica
Arabidopsis suecica
Alchemilla oxyodonta
Saxifraga osloensis
Calamagrostis chalybaea

Of these, three are allopolyploids: *Corydalis gotlandica* (Liden 1991), *Arabidopsis suecica* (Hylander 1947; Løve 1961), and *Saxifraga osloensis* (Knaben 1954)). *Alchemilla oxydonta* and *Calamagrostis chalybaea* are apomicts.

B British Isles endemics

Vascular plants

I Montane-alpine endemics

   1 Scottish endemics

Athyrium flexile
Cerastium scoticum
C. arcticum ssp. *edmondstonii*
Cochlearia micacea
Ranunculus flammula ssp. *scoticus* (also in the Faeroes)
Gentianella amarella ssp. *druceana*
   ssp. *septentrionalis*
Primula scotica
Euphrasia rhumica
E. foulaensis
E. heslop-harrisonii
E. campbelliae
E. marschallii
Artemisia norvegica var. *scotica*

239

Calamagrostis scotica
Dactylorhiza maculata ssp. *rhoumensis*
D. majalis ssp. *cambrensis* (also in England)

It is a striking correlation or coincidence that the distribution of *Primula scotica* in Caithness and the Orkneys is strictly limited to the area considered by the author to have been ice-free during the Weichselian maximum.

2 Irish endemics

Arenaria ciliata ssp. *hibernica*
Arabis hirsuta ssp. *brownii*
Saxifraga hartii
Gentianella amarella ssp. *hibernica*
Euphrasia salisburgensis var. *hibernica*

In addition *Sisyrinchium bermudianum* is represented by a race with a different chromosome number from its American counterpart.

3 Scottish–Irish endemics

Ranunculus flammula ssp. *minimus*
Cochlearia scotica
Dactylorhiza fuchsii ssp. *okellyi*
D. fuchsii ssp. *hebridensis*
D. majalis ssp. *occidentalis*

In addition *Eriocaulon aquaticum* is represented by a race with a different chromosome number from its American counterpart.

4 Welsh endemics

Euphrasia cambrica
E. rivularis (also in Lake District)

5 Pennines–Lake District endemics

Arenaria norvegica ssp. *anglica*
Helianthemum levigatum
Alchemilla minima

II Coastal endemics

Rumex acetosa ssp. *hibernica*
Rhynchosinapis wrightii

R. monensis
Senecio integrifolius ssp. *maritimus*
Catabrosa aquatica ssp. *minor*
Spartina anglica
Epipactis dunensis
Dactylorhiza incarnata ssp. *coccinea*

Of these *Spartina anglica* is a recent allopolyploid arising from a crossing between *S. maritima* and the American *S. alterniflora*.

III Other (southern lowland) endemics

Scleranthus perennis ssp. *prostratus*
Fumaria capreolata ssp. *babingtonii*
F. occidentalis
F. purpurea (also in Ireland)
F. muralis ssp. *neglecta*
Anthyllis vulneraria ssp. *corbieri*
Geranium purpureum ssp. *forsteri*
Gentianella anglica ssp. *anglica*
    ssp. *cornubiensis*

Linum perenne ssp. *anglicum*
Euphrasia vigursii
E. pseudokerneri (also in Ireland)
E. anglica (?)
Senecio cambrensis
Bromus interruptus
B. pseudosecalinus

Of these *Fumaria occidentalis* is considered an allopolyploid from a cross between *F. capreolata* and *F. bastardii* (Stace 1975). *F. occidentalis* is confined to artificial habitats. *Linum perenne* ssp. *anglicum* is a tetraploid related to the widespread *L. perenne*. *Senecio cambrensis* is a recent allopolyploid from a cross between *S. squalidus* and *S. vulgaris*. *Euphrasia vigursii* is considered a stabilised hybrid from *E. anglica* and *E. micrantha* (Yeo in Stace 1975). *Bromus interruptus* is of doubtful origin appearing as a weed in fields of cultivation of *Onobrychis*. *Bromus pseudosecalinus* is a diploid related to the widespread, tetraploid *B. secalinus* of uncertain origin.

Poikilohydric endemics

With indications of endemic area. BI, British Isles; I, Ireland; Sc, Scotland, W, Wales; E, England.

Liverworts

Cephaloziella nicholsonii BI (southern lowlands)
Fossombronia fimbriata I, Sc, W
Plagiochila atlantica I (also in Finistère)

P. britannica   I, Sc (also in Finistère)
Teleranea murphyae   BI (southern lowlands)

Mosses

Amblystegium serpens var. *salinum*   BI (coastal)
Barbula mammilosa   Sc
B. maxima   I
B. nicholsonii   BI
B. tomaculosa   BI (southern lowlands)
Brachythecium appleyardii   BI (southern lowlands)
Bryoerythrophyllum caledonium   Sc
Bryum capillare var. *rufifolium*   I, Sc
B. dixonii   Sc
B. lawersianum   Sc
Campylopus atrovirens var. *gracilis*   I, Sc (southern lowlands)
Ctenidium molluscum var. *fastigiatum*   I, E
Ditrichum cornubicum   BI (southern lowlands)
D. plumbicola   BI (southern lowlands)
D. zonatum var. *scabrifolium*   I, Sc
Fissidens celticus   BI
Fontinalis antipyretica var. *cymbifolia*   I, E
F. squamosa var. *dixonii*   Sc
Grimmia retracta   BI
G. trichophylla var. *stirtonii*   I, E
Gymnostomum insigne   I, Sc
Pictus scoticus   Sc
Pohlia scotica   Sc
Sphagnum skyense   Sc
Thamnobryum angustifolium   E
Tortella limosella   Sc, not seen since 1906
Weissia mittenii   BI (southern lowlands)
W. multicapsularis   BI (southern lowlands)

C North Atlantic endemics in the area of British Isles, Fennoscandia, the Faeroes and Iceland

Vascular plants

Dactylorhiza purpurella
Arenaria norvegica
Ranunculus flammula ssp. *scoticus*
Papaver radicatum
Alchemilla faeroensis
Anthyllis vulneraria ssp. *lapponica*
Angelica archangelica ssp. *litoralis*
Euphrasia arctica
E. scottica
E. confusa
E. foulaensis
E. ostenfeldii

Liverworts

Herbertus aduncus ssp. *hutchinsiae*
H. borealis
Lepidozia pearsonii
Plagiochila norvegica

Mosses

Anoectangium warburgii
Campylopus atrovirens var. *falcatus*
Grimmia trichophylla var. *stirtonii*
Isothecium myosuroides var. *brachythecioides*
Oxystegus tenuirostris var. *holtii*
Schistidium apocarpum var. *homodictyon*
Trematodon laetevirens
Weissia perssonii

Lichens

Anaptychia ciliaris ssp. *mamillata*
Leptogium britannicum
Parmeliopsis esorediata
Pilophorus strumaticus
Stereocaulon subdenudatum

Extra-European disjunctions –
bryophytes and lichens

1 Macaronesian disjuncts

Liverworts

Acrobolbus wilsonii
Adelanthus decipiens
Cephalozia hibernica
*Jubula hutchinsiae*ssp. *hutchinsiae*
*Lejeunea flava*ssp. *moorei*
L. hibernica
L. holtii
L. mandonii
Leptoscyphus cuneifolius
Metzgeria leptoneura
Microlejeunea hibernica
Plagiochila killarniensis
P. spinulosa
Radula carringtonii
R. holtii
Telaranea nematodes

Mosses

Campylopus shawii
Cyclodictyon laetevirens
Glyphomitrium daviesii
Heterocladium heteropterum var. *heteropterum*
Myurium hochstetteri
Rhacomitrium ellipticum

Lichens

Cladonia cyathomorpha
C. stereoclada
Degelia ligulata
Polychidium dendriscum
Pseudocyphellaria lacerata
P. norvegica
Usnea subscabrosa

2 Southeast Asiatic disjuncts

Liverworts

Anastrophyllum joergensenii
Eremonotus myriocarpus
Lepidozia cupressina
Lophozia decolorans
Plagiochila (*Jamsoniella*) *carringtonii*
Scapania nimbosa
S. ornithopodioides

Mosses

Bartramidula wilsonii
Campylopus subulatus
Dicranodontium subporodictyon
Rhacomitrium himalayanum

Lichens

Cetrelia olivetorum

3 Pacific American disjuncts

Liverworts

Herbertus aduncus
H. stramineus
Kurzia trichoclados
Marsupella boeckii var. *stableri*
M. commutata

Mosses

Dicranum tauricum
Encalypta longicollis
E. spathulata
Gymnostomum insigne
Hyophila stanfordensis
Leptodontium recurvifolium
Plagiothecium undulatum
Schistidium trichodon
Tortula amplexa

Lichens

Bryoria fremontii
B. tortuosa
Cladonia portentosa
C. umbricola
Dermatocarpon arnoldianum
Leptogium corniculatum
Letharia vulpina
Parmelia loxodes
Peltigera britannica
Placynthium subradiatum

References

Aas, B. (1964). Bjørke- og barskogsgrenser i Norge. M.sci. thesis, University of Oslo.

Aas, B. & Faarlund, T. (1988). Postglasiale skoggrenser i sentrale sørnorske fjelltrakter. 14C datering av subfossile furu- og bjørkerester. *Norsk Geogr. Tidsskr.* **42**: 25–61.

Abrahamsen, J., Dahl, E., Jacobsen, N. K., Kalliola, R., Påhlsson, L. & Wilborg, L. (1977). Naturgeografisk regionindelning av Norden. *Nordiska utredningar 34.* Nordiska ministerrådet.

Ahlner, S. (1948). Utbredningstyper bland nordiska barrträdslavar. *Acta Phytogeographica Suecica* **22**: 1–257.

Ahti, T. (1977). Lichens of the boreal coniferous zone. In Seward, M. R. D. (ed.): *Lichen Ecology*, pp. 145–181. Academic Press, London.

Alexandrov, V. Ya. (1977). *Cells, molecules and temperature.* Ecological Studies 21. Springer-Verlag, Berlin.

Alm, T. & Birks, H. H. (1991). Late Weichselian flora and vegetation of Andøya, northern Norway – macrofossil (seed and fruit) evidence from Nedre Æråsvatn. *Nordic Journal of Botany* **11**: 465–76.

Ammann, K. & Ammann, B. (1969). Die fennoskandische Verbreitung von *Pilophorus* (Tuck.) Th. Fr., Stereocaulaceae. *Herzogia* **1**: 87–94.

Andersen, B. G. (1954). Randmorener i Sørvest-Norge. *Norsk Geogr. Tidsskr.* **14**: 273–342.

Andersen, B. G. (1981). Late Weichselian ice sheets in Eurasia and Greenland. In Denton, G. H. & Hughes, T. J. (eds.): *The last great ice sheets*, pp. 1–65. John Wiley, New York.

Anonymous (1973). Tables of temperature, relative humidity, precipitation and sunshine for the world. Part III. Europe and the Azores. *Met. O. 856c.* Her Majesty's Stationary Office, London.

Anonymous (1983). List of rare, threatened and endemic plants in Europe. *Council of Europe. Nature and environment series 27,* 357 pp.

Arnell, S. (1956). *Illustrated moss flora of Fennoscandia. II. Hepaticae.* Lund, Gleerup.

Arwidsson, T. (1943). Studien über de Gefässpflanzen in de Hochgebirgen der Pite Lappmark. *Acta Phytogeographica Suecica* **17**: 1–274.

Ballantyne, C. K. (1990). The late Quaternary glacial history of the Trotternish Escarpment, Isle of Skye, and its implications for ice-sheet reconstruction. *Proceedings of the Geological Association* **101**: 171–86.

Ballantyne, C. K. (1994). Gibbsitic soils on former nunataks: implications for ice sheet reconstruction. *Journal of Quaternary Science* **9**: 73–80.

Ballantyne, C. K. & McCarroll, D. (1995). The vertical dimensions of Late Devensian glaciation on the mountains of Harris and southern Lewis, Outer Hebrides, Scotland. *Journal of Quaternary Science* **10**: 211–23.

Bay, C. (1992). A phytogeographical study of the vascular plants of northern Greenland – north of 74° northern latitude. *Medd. Grønland. Bioscience* **36**, 102 pp.

Becherer, A. (1956). Florae Vallesiacae Supplementum. Supplement zu Henri Jaccards Catalogue de la Flore Vallaisanne. *Denkschr. Schweizerischen Naturforschenden Ges.* **81**: 1–557.

Behre, K.-E. (1988). The role of man in European vegetation history. In Huntley, B. & Webb, T. III (eds.): *Vegetation history. Handbook of Vegetation Science 7*, pp. 633–72. Kluwer, Dordrecht.

Bennett, K. D. (1984). The post-glacial history of *Pinus sylvestris* in the British Isles. *Quaternary Science Reviews* **3**: 133–55.

Bennett, K. D., Tzedakis, P. C. & Willis, K. J. (1991): Quaternary refugia of North European trees. *Journal of Biogeography* **18**: 103–15.

Bennike, O. (1987). News from the Plio-Pleistocene Kap København Formation, North Greenland. *Polar Research* **5**: 339–40.

Berg, R. (1963). Disjunksjoner i Norges fjellflora og de teorier som er fremsatt til forklaring av dem. *Blyttia* **21**: 133–77.

Berglund, B. E. (1966a). Late Quaternary vegetation in eastern Blekinge, South-Eastern Sweden. A pollen-analytical study. I. Late-glacial time. *Opera Botanica* **12** (1): 1–180.

Berglund, B. E. (1966b). Late Quaternary vegetation in eastern Blekinge, South-Eastern Sweden. A pollen-analytical study. II. Post-glacial time. *Opera Botanica* **12** (2): 1–190.

Berglund, B. E. (1979). The deglaciation of southern Sweden 13 500–10 000 B.P. *Boreas* **8**: 89–118.

Biebl, R. (1965). Temperaturresistenz tropischer Pflanzen auf Puerto Rico. *Protoplasma* **59**: 133–56.

Billings, W. D. & Mooney, H. A. (1968). The ecology of arctic and alpine plants. *Biological Reviews* **43**: 481–530.

Birks, H. H. (1994). Plant macrofossils and the nunatak theory of per-glacial survival. *Dissertationes Botanicae* **234**: 129–43.

Birks, H. J. B. (1976). The distribution of European pteridophytes: a numerical analysis. *New Phytologist* **77**: 257–87.

Birks, H. J. B. (1986). Late-Quaternary biotic changes in terrestrial and lacustrine environments, with particular reference to north-west Europe. In Berglund, B. E. (ed.): *Handbook of Holocene Palaeoecology and Palaeohydrology*, pp. 3–65. John Wiley, Chichester.

Birks, H. J. B. (1987). Recent methodological developments in quantitative descriptive biogeography. *Annales Zoologici Fennici* **24**: 165–78.

Birks, H. J. B. (1988). Long-term ecological change in the British uplands. In Usher, M. B. & Thompson, D. B. A. (eds.): *Ecological Change in the Uplands*. pp. 32–7. Special Publ. Number 7 of the British Ecological Soc. Blackwell Science Publications, Oxford.

Birks, H. J. B. (1989). Holocene isochrone maps and patterns of tree-spreading in the British Isles. *Journal of Biogeography* **16**: 503–40.

Birks, H. J. B. (1990). Changes in vegetation and climate during the Holocene of Europe. In Boer, M. M. & De Groot, R. S. (eds.): *Landscape–ecological impact of climatic change*. Proc. Eur. Conf., Lunteren, The Netherlands, pp. 133–57. IOS Press, Amsterdam.

Birks, H. J. B. (1991). Floristics and vegetational history of the Outer Hebrides. In Pankhurst, R. J. & Mullin. J. M. (eds.): *Flora of the Outer Hebrides*, pp. 32–37. Natural History Museum Publ., London.

Birks, H. J. B. (1993). Is the hypothesis of survival on glacial nunataks necessary to explain the present-day distributions of Norwegian mountain plants? *Phytocoenologia* **23**: 399–426.

Björkman, O. & Holmgren, P. (1961). Studies of climatic ecotypes of higher plant respiration in different populations of *Solidago virgaurea*. *Kungl. Lantbrukshögskolans Annaler* **27**: 297–301.

Blytt, A. (1869). *Om vegetationforholdene ved Sognefjorden*. Christiania.

Blytt, A. (1876). *Essay on the immigration of the Norwegian flora during alternating rainy and dry periods*. Christiania.

Borgen, L. (1987). Postglasial evolusjon i Nordens flora – en oppsummering. *Blyttia* **45**: 147–87.

Bowen, D. Q. & Sykes, G. A. (1988). Correlation of marine events and glaciation on the northeast Atlantic margin. *Philosophical Transactions of the Royal Society of London, B* **318**: 619–35.

Braun-Blanquet, J. (1923). *L'origine et développement des flores dans le Massif Centrale de France*. Paris and Zurich.

Braun-Blanquet, J. & Rübel, E. (1932–5). Flora von Graubünden. *Veröff. Geobot. Inst. Rübel, Zürich, H 7*, 1695 pp.

Brochmann, C. (1992). Polyploid evolution in arctic-alpine *Draba* (Brassicacae). *Sommerfeltia suppl.* **4**: 1–37.

Brochmann, C., Soltis, P. S. & Soltis, D. E. (1992). Multiple origins of the octoploid endemic *Draba cacuminum*. Electrophoretical and morphological evidence. *Nordic Journal of Botany* **12**: 257–72.

Brockmann-Jerosch, H. (1919). Baumgrenze und Klimacharacter. *Ber. Schw. Bot. Ges.* **24**: 1–255.

Brockmann-Jerosch, H. (1923). Betrachtungen über Pflanzenausbreitung. *Verh. Naturforsch. Ges. Basel* **35**: 382–404.

Bronger, C. (1992). *Silene armeria* in Norway. *Blyttia* **50**: 1–11. [Norwegian with English summary.]

Bøcher, T. W. (1938). Biological distribution types in the flora of Greenland. *Medd. Grønland* **106**: 1–339.

Bøcher, T. W. (1951). Distribution of plants in the circumpolar areas in relation to ecological and historical factors. *Journal of Ecology* **39**: 376–95.

Bøcher, T. W. (1963). Experimental and cytological studies on plant species. VIII. Racial differentiation in amphi-atlantic *Viscaria alpina*. *Biol. Skr.* **11**, 6: 1–33.

Bøcher, T. W. (1975). *Det grønne Grønland.* Rhodos Forlag, Copenhagen.

Bøcher, T. W., Holmen, K. & Jacobsen, K. (1968). *The flora of Greenland.* Haase & Søn, Copenhagen.

Bøcher, T. W., Fredskild, B., Holmen, K. & Jacobsen, K. (1978). *Grønlands Flora* 3, rev. edn. Haase & Søn, Copenhagen.

Canada Soil Survey Committee (1978). *The Canadian system of soil classification.* Canadian Department of Agriculture Publication 1646.

Clapham, A. R., Tutin, T. G. & Moore, D. M. (1987). *Flora of the British Isles*, 3rd edn. Cambridge University Press, Cambridge.

Clausen, E. (1952). Hepatics and humidity; a study of Hepatics in a British tract and the influence of relative humidity on their distribution. *Dansk Bot. Ark.* **15**, –80.

CLIMAP Project Members (1976). The surface of the ice-age earth. *Science* **191**: 1131–7.

Conolly, A. P. & Dahl, E. (1970). Maximum summer temperature in relation to modern and Quaternary distributions of certain arctic-montane species in the British Isles. In Walker, D. & West, R. G. (eds.): *Studies in the vegetational history of the British Isles.* pp. 159–223. Cambridge University Press, Cambridge.

Coope, G. R., Morgan, A. & Osborne, P. J. (1971). Fossil Coleoptera as indicators of climatic fluctuations during Last Glaciation in Britain. *Palaeogeography, Palaeoclimatology, Palaeoecology* **10**: 87–101.

Coxon, P. & Waldren, S. (1995). The floristic record of Ireland's Pleistocene temperate stages. In Preece, R. C. (ed.): *Island Britain: a Quaternary perspective*, pp. 243–67. Geological Society Special Publication 96, London.

Dahl, E. (1950). Studies in the macrolichenflora of South West Greenland. *Medd. Grønland* **150** (2), 1–176.

Dahl, E. (1951). On the relation between summer temperature and the distribution of alpine vascular plants in the lowlands of Fennoscandia. *Oikos* **3**: 22–52.

Dahl, E. (1954). Weathered gneisses at the island of Runde, Sunmøre, Western Norway and their geological interpretation. *Nytt Magasin Bot.* **3**: 5–23.

Dahl, E. (1955). Biogeographic and geologic indications of unglaciated areas in Scandinavia during the glacial ages. *Bulletin of the Geological Society of America* **66**: 1499–1520.

Dahl, E. (1957). Rondane mountain vegetation in South Norway and its relation to the environment. *Skr. Det Norske Vidsk.- Akad. Oslo. I Mat.- Naturv. Kl. 1956,* **3**: 1–373.

Dahl, E. (1961). Refugieproblemet og de kvartærgeologiske metodene. *Svensk Naturvetenskap* **14**: 81–96.

Dahl, E. (1963a). On the heat exchange of a wet vegetation surface and the ecology of *Koenigia islandica. Oikos* **14**: 190–211.

Dahl, E. (1963b). Plant migrations across the North Atlantic Ocean and their importance for the paleogeography of the region. In Løve, A. & Løve, D. (eds.): *North Atlantic biota and their history,* pp. 173–88. Pergamon Press, London.

Dahl, E. (1964). Present-day distribution of plants and past climate. In Hester, J. J. & Schoenwetter, J. (eds.): *The reconstruction of past environments,* pp. 52–60. Fort Burgwin research centre, New Mexico, no. 3.

Dahl, E. (1966). Plantenes varmeveksling med omgivelsene og dens betydning for plantenes morfologi og utbredelse. (The heat exchange of plants and its importance to plant morphology and distribution.) *Blyttia* **24**: 105–29 [Norwegian with English summary.]

Dahl, E. (1986). Zonation in Arctic and Alpine tundra and fellfield ecobiomes. In Polunin N. (ed.): *Ecosystem theory and application,* pp. 35–62. John Wiley, New York.

Dahl, E. (1987). The nunatak theory reconsidered. *Ecological Bulletin* **38**: 77–94.

Dahl, E. (1989). Nunatakk-teorien II. Endemismeproblemet. (The nunatak theory II – The problem of endemism). *Blyttia* **47**: 163–74. [Norwegian with English summary.]

Dahl, E. (1990a). Probable effects of climatic change due to the greenhouse effect on plant productivity and survival in North Europe. *Norsk Inst. Natur-forskning (NINA) Notat 004*: 7–18.

Dahl, E. (1990b). History of the Scandinavian alpine flora. In Gjærevoll, O. (ed.): *Maps of distribution of Norwegian vascular plants. II. Alpine plants,* pp. 16–21. Tapir, Trondheim.

Dahl, E. (1991). Nunatakkteorien III. Amfiatlanter og disjunkter. (The nunatac theory III – amphiatlants and disjuncts). *Blyttia* **49**: 17–33. [Norwegian with English summary.]

Dahl, E. (1992a). Nunatakkteori. IV. Hvor fantes isfrie områder og hva slags planter kunne leve på dem? (The nunatac theory IV – Which areas remained unglaciated during the Pleistocene glacial ages and which plants were able to survive?) *Blyttia* **50**: 23–35. [Norwegian with English summary.]

Dahl, E. (1992b). Relations between macro-meteorological factors and the distribution of vascular plants in northern Europe. *Rapport bot. ser. 1992, 1*: 31–59. Univ. Trondheim Vidensk. museet. Trondheim.

Dahl, E. & Mork, E. (1959). Om sambandet mellom temperatur, ånding og vekst hos gran (*Picea abies* (L.) Karst.). (On the relationships between temperature, respiration and growth in Norway spruce (*Picea abies* (L.) Karst.)). *Medd. Det Norske Skogforsøksvesen* **53**: 82–93. [Norwegian with English summary.]

Dahl, E., Elven, R., Moen, A. & Skogen, A. (1986). *Vegetasjonsregioner. Kart 1: 500000. Nationalatlas for Norge Kartblad 4.1.1.*

Dalla Torre, F. W. & Sarntheim, L. G. (1906–12). *Die Farn- und Blutenpflanzen (Pteridophyta et Siphonogama) von Tirol, Vorarlberg und Liechtenstein.* 1 Teil 1906, 2 Teil 1909, 3 Teil 1912. Innsbruck.

Daniels, R. E. & Eddy, A. (1990). *Handbook of the European Sphagna.* Institute of Terrestrial Ecology. Her Majesty's Stationery Office, London.

Daubenmire, R. (1954). Alpine timberlines in the Americas and their interpretation. *Butler University Botanical Studies 11*: 119–36.

Degelius, G. (1935). Das ozeanische Element der Strauch- und Laubflechtenflora von Skandinavien. *Acta Phytogeographica Suecica* **7**: 1–411.

Denko, E. I., Kislyuk, I. M. & Shukhtina, H. G. (1981). Primary thermostability of cells and a problem of adaptation of plants to conditions of cold climate. *Flora* **171**: 419–52.

Denton, G. H. & Hughes, T. J. (eds.) (1981). *The last great ice sheets.* John Wiley, New York.

de Vries, F. W. T. P. (1974). Substrate utilization and respiration in relation to growth and maintenance in higher plants. *Netherlands Journal of Agricultural Science* **22**: 40–4.

de Vries, F. W. T. P. (1975). Use of assimilates in higher plants. In Cooper, J. P. (ed.): *Photosynthesis and productivity in different environments.* International Biological Programme **5**: 459–80. IBP, Cambridge.

de Vries, F. W. T. P., Jansen, D. M., ten Berge, H. F. M. & Bakema, A. (1989). Simulation of eco-physiological processes of growth in several annual crops. *Simulation Monographs 29.* Pudoc, Wageningen.

Di Castri, F., Hansen, A. J. & Debussche, M. (eds.) (1987). *Biological invasions in Europe and the Mediterranean Basin.* Kluwer, Dordrecht.

Dickson, J. H. (1992). Some recent additions to the Quaternary flora of Scotland and their phytogeographical, palaeoclimatic and ethnobotanical significance. A review. *Acta Botanica Fennica* **144**: 51–7.

Dilks, T. J. K. & Proctor, M. C. F. (1974). The pattern of recovery of bryophytes after desiccation. *Journal of Bryology* **8**: 97–115.

Dilks, T. J. K. & Proctor, M. C. F. (1976a). Seasonal variation in desiccation tolerance in some British bryophytes. *Journal of Bryology* **9**: 239–47.

Dilks, T. J. K. & Proctor, M. C. F. (1976b). Effects of intermittent desiccation on bryophytes. *Jourrnal of Bryology* **9**: 249–64.

Dupont, P. (1990). Atlas partiel de la flora de France. *Collection Patrimoines Naturels* **3**: 1–441. Museum National d'Histoire Naturelle, Paris.

DuRietz, G. E. (1940). Problems of bipolar plant distribution. *Acta Phytogeographica Suecica* **13**: 215–82.

Edwards, M. E. (1986). Disturbance histories of four Snowdonian woodlands and their relation to Atlantic bryophyte distributions. *Biological conservation* **37**: 301–20.

Eide, E. (1932). Furuens vekst og foryngelse i Finmark. *Medd. Det Norske Skogforsøksvesen* **15**: 329–429.

Eldholm, O. & Thiede, J. (1986a). Formation of the Norwegian Sea from the leg 104 shipboard party. *Nature* **319***: 360–1.

Eldholm, O. & Thiede, J. (1986b). Above the Arctic Circle. Reflector identified, glacial onset seen. *Grotimes March 1986*: 12–15.

Ellenberg, H. (1978). *Die Vegetation Mitteleuropas mit den Alpen in ökologischer Sicht*, 2nd edn. Ulmer Verlag, Stuttgart.

Ellenberg, H., Weber, H. E., Dull, R., Wirth, V., Werner, W. & Paulissen, D. (1991). Zeigerwerte der Gefasspflanzen in Mitteleuropa. *Scripta Geobotanica* **18**: 1–248.

Ericsson, L. & Wallentius, H.-G. (1979). Sea-shore vegetation around the Gulf of Bothnia. *Wahlenbergia* **5**: 1–118.

Esseen, P.-A., Ericson, L., Lindström, H. & Zackrisson, O. (1981). Occurrence and ecology of *Usnea longissima* in Central Sweden. *Lichenologist* **13**: 177–90.

FAO-UNESCO (1981). *Soil map of the world. 1:5 000 000. V Europe*. 199 pp., 10 maps.

Farvager, C. (1972). Endemism in the montane floras of Europe. In Valentine, D. H. (ed.): *Taxonomy, phytogeography and evolution.* pp. 191–204. Academic Press, London.

Feilberg, J. (1984). A phytogeographical study of South Greenland. *Medd. Grønland. Bioscience* **15**: 1–71.

Fernald, M. L. (1925). Persistence of plants in unglaciated areas of boreal America. *Memoirs of the American Academy of Science* **15**(3): 237–342.

Florin, M.-B. (1979). The younger Dryas vegetation at Kolmården in southern central Sweden. *Boreas* **8**: 145–52.

Florschutz, F. (1958). Steppen- und Salzwiesenelemente aus der Flora der Letzten und Vorletzten Eiszeit in den Niederlanden. *Flora* **146**: 489–92.

Forman, R. T. T. (1964). Growth under controlled conditions to explain the hierarchical distributions of a moss, *Tetraphis pellucida*. *Ecological Monographs* **34**: 1–25.

Forman, S. L. & Miller, G. H. (1984). Time-dependent soil morphologies and pedogenic processes on raised beaches, Brøggerhalvøya, Spitsbergen, Svalbard Archipelago. *Arctic and Alpine Research* **16**: 381–94.

Foss, P. J. & Doyle, G. J. (1990). The history of *Erica erigena* R. Ross, an Irish plant with a disjunct European distribution. *Journal of Quaternary Science* **5**: 1–16.

Franz, H. (1979). *Ökologie der Hochgebirge*. Ulmer Verlag, Stuttgart.

Frenzel, B. (1960). Die Vegetations und Landschaftzonen Nordeurasiens während der letzten Eiszeit und während der postglazialen Warmezeit. *Abh. mat.-naturv. Kl. Akad. Wiss. Litt., Mainz 1959*, **13**: 934–1099.

Frenzel, B. (1968). Grundzüge der pleistozänen Vegetationsgeschichte Nord-Eurasiens. *Erdwiss. Fortschr. 1.* Wiesbaden.

Frenzel, B. (1987). Grundprobleme der Vegetationsgeschichte Mitteleuropas während des Eiszeitalters. *Mitt. Naturforsch. Ges. Luzern* **29**: 99–122.

Fries, T. C. E. (1913). *Botanische Untersuchungen im nordlichsten Schweden.* Akad. Avhandl., Uppsala & Stockholm, 361 pp.

Funder, S. (1989). Quaternary geology of the ice-free areas and adjacent shelves of Greenland. In Fulton, R. J. (ed.): *Quaternary geology of Canada and Greenland,* pp. 743–92. Geology of Canada. No. 1. Canadian Government Publishing Centre, Ottawa.

Funder, S., Abrahamsen, N., Bennike, O. & Feyling-Hanssen, R. W. (1985). Forested Arctic: evidence from Greenland. *Geology* **13**: 542–6.

Fægri, K. (1950). Studies on the Pleistocene in West Norway. On the immigration of Picea abies (L.) Karst. *Univ. Bergen Årbok 1949 Naturvitensk. Rekke* **1**: 1–52.

Fægri, K. (1958–60). *Norges planter, I–II.* Cappelen, Oslo.

Fægri, K. (1960). *Map of distribution of Norwegian plants. I. The coast plants.* Oslo University Press, Oslo.

Faarlund, T. & Aas, B. (1991). Mountain forests in southern Norway during the post-glacial period. *Viking 1991*: 113–37.

Galloway, D. J. (1985). *Flora of New Zealand. Lichens.* Hasselberg, Government Printer, Wellington.

Gamisans, J. (1976–78). La végétation des montagnes corses. *Phytocoenologia* **3**: 425–98 1976, **4**: 35–131, 133–79, 317–76, 377–432, 1978.

Gates, D. M. (1980). *Biophysical ecology.* Springer-Verlag, Berlin.

Gauch, H. G. (1982). *Multivariate analysis in community ecology.* Cambridge University Press, Cambridge.

Gauslaa, Y. (1984). Heat resistance and energy budget in different Scandinavian plants. *Holarctic Ecology* **7**: 1–78.

Gauslaa, Y. (1985). The ecology of Lobarion pulmonariae and Parmelion caperatae in *Quercus*-dominated forests in south-west Norway. *Lichenologist* **17**: 117–40.

Gelting, P. (1934). Studies on the vascular plants of East Greenland between Franz Joseph Fjord and Dove Bay. *Medd. Grønland 116*, **3**: 1–340.

Gentry, A. (1986). Endemism in tropical versus temperate plant communities. In Soulé, M. E. (ed.): *Conservation biology. The science of scarcity and diversity,* pp. 153–181. Sinauer Associates, Sunderland, Mass.

George, M. F., Burke, M. J., Pellett, H. M. & Johnson, G. (1974). Low temperature exotherms and woody plant distribution. *Hort. Science* **9**: 519–22.

Gjærevoll, O. (1973). *Plantegeografi.* Trondheim, Oslo, Bergen, Tromsø. 173 pp.

Gjærevoll, O. (1990). *Maps of distribution of Norwegian vascular plants. II. Alpine plants.* Tapir, Trondheim.

Gjærevoll, O. (1992). *Plantegeografi.* Tapir, Trondheim.

Gjærevoll, O. & Ryvarden, L. (1977). Botanical investigations on the J. A. D. Jensens Nunatakker in Greenland. *Det K. Norske Vidensk. Selsk. skr. 40,* **4**: 1–40.

Godwin, H. (1975). *History of the British flora. A factual basis for phytogeography.* 2nd edn. Cambridge University Press, Cambridge.

Grace, J. B. & Tilman, D. (1990). *Perspectives on plant competition.* Academic Press, New York.

Greig-Smith, P. (1950). Evidence from the hepatics on the history of the British flora. *Journal of Ecology* **38**: 320–44.

Gries, D. (1991). Die ökologische Bedeutung der Eisen-Anneigungsfahigkeit bei Poaceen. *Diss. Mat.- Naturw. Fachberichte Universität Göttingen*, 85 pp., Anhang.

Haeupler, H. & Schönfelder, P. (1988). *Atlas der Farn und Blutenpflanzen der Bundesrepublik Deutschland.* Ulmer Verlag, Stuttgart.

Hafsten, U. (1956). Pollen-analytic investigations on the late Quaternary development in the Oslofjord area. *Univ. Bergen, Årbok 1956. Naturvitensk. Rekke* **4**: 1–161.

Hafsten, U. (1991). The history of the spruce forest in Norway under exposure. *Blyttia* **49**: 171–81.

Hagem, O. (1917). Furuens og granens frøsætting i Norge. *Medd. Vestlandets Forstlige Forsøksstasjon* **1**, 2: 1–188.

Hagem, O. (1947). The dry matter increase of coniferous seedlings in winter. *Medd. Vestlandets Forstlige Forsøksstasjon* **26**: 1–317.

Hantke, R. (1978). *Eiszeitalter. Bd. 1. Die jüngste Erdgeschichte der Schweitz und ihrer Nachbargebiete. Klima, Flora, Mensh Alt- und Mittel-Pleistozan Vogesen, Schwarzwald, Schwabische Alb Adelegg.* Ott Verlag AG, Thun.

Haraldsen, K. B. & Wesenberg, J. (1993). Population genetic analyses of an amphi-atlantic species: *Lychnis alpina* (Caryophyllaceae). *Nordic Journal of Botany* **13**: 377–82.

Hengeveld, R. (1990). *Dynamic biogeography.* Cambridge University Press, Cambridge.

Hermes, K. (1955). Die Lage der oberen Waldgrenze in den Gebirgen der Erde und ihr Abstand zur Schneegrenze. *Kölner geogr. Arb.* **5**: 1–277.

Heslop-Harrison, J. (1973). The North American and Lusitanian elements in the flora of the British Isles. In Lousley, J. E. (ed.): *The changing flora of Britain.* pp. 105–23. Arbroath.

Hilbig, W. (1982). Preservation of agrestal weeds. In Holzner, W. & Numata, E. (eds.): *Biology and ecology of Weeds. Geobotany 2*, pp. 57–9. Junk, The Hague.

Hill, M. O., Preston, C. D. & Smith, A. J. E. (1991–4). *Atlas of the Bryophytes of Britain and Ireland. I-III.* Harley Books, Colchester.

Hillier, S. H., Walton, D. W. H. & Wells, D. A. (eds.) (1990). *Calcareous grasslands – ecology and management.* Bluntisham Books, Bluntisham, UK.

Hinttikka, V. (1963). Über das Grossklima einiger Pflanzenareale in zwei Klimakoordinaten dargestellt. *Ann. Bot. Soc. Zool.-bot. Fenn. Vanamo 34*, 5: 1–64.

Hjelmqvist, H. (1955). Die älteste Geschichte der Kulturpflanzen in Schweden. *Opera Botanica* **1**(34): 1–186.

Holmboe, J. (1925). Einige Grundzüge der Pflanzengeographie Norwegens. *Bergens Museums Aarbok* 1924–25, **3**:1–54.

Holzner, W. & Numata, E. (eds.) (1982). *Biology and ecology of weeds. Geobotany 2.* Junk, The Hague.

Hultén, E. (1937). *Outline of the history of arctic and boreal biota during the Quaternary period.* Thule Förlag, Stockholm.

Hultén, E. (1950). *Atlas of the distribution of vascular plants in NW. Europe,* 1st edn. Generalstabens litogr. anst. Förlag, Stockholm.

Hultén, E. (1955). The isolation of the Scandinavian mountain flora. *Acta Sococietas Fauna et Flora Fennica* **72**(8): 1–22.

Hultén, E. (1958). The Amphi-Atlantic plants and their phytogeographic connections. *Kungl. Svenska Vetensk.-akad. Handl. Ser. 4b,* **7**: 1–340.

Hultén, E. (1962). The circumpolar plants I. *Kungl. Svenska Vetensk.-akad. Handl. Ser. 4,* **8**(5): 1–275.

Hultén, E. (1971a). *Atlas of the distribution of vascular plants in northwestern Europe.* 2nd edn. AB Kartografiska inst. Stockholm.

Hultén, E. (1971b). The circumpolar plants II. *Kungl. Svenska Vetensk.- akad. Handl. Ser. 4,* **13**(1): 1–463.

Hultén, E. & Fries, M. (1986). *Atlas of the North European vascular plants north of the Tropic of Cancer.* I–II. Koeltz, Koenigstein.

Huntley, B. (1988). Glacial and Holocene vegetation history – 20 ky to present. Europe. In Huntley, B. & Webb, T. III (eds.): *Vegetation history. Handbook of vegetation science 7,* pp. 339–83. Kluwer, Dordrecht.

Huntley, B. & Birks, H. J. B. (1983). *An atlas of past and present pollen maps for Europe 0–13 000 years ago.* Cambridge University Press, Cambridge.

Hustich, I. (1966). On the forest-tundra and the northern tree-lines. *Rep. Kevo Subarctic Sta.* **3**: 7–47.

Hutchinson, G. E. (1957). Concluding remarks. *Cold Spring Harbor Symposia on Quantitative Biology* **22**: 415–27.

Hylander, N. (1947). *Cardaminopsis suecica* (Fr.) Hiit, a northern amphiploid species. *Bulletin Jardin Botanique, Bruxelles* **27**: 591–604.

Imhof, E. (1900). Die Waldgrenze in der Schweitz. *Beitr. Geophysik. Zeitschr. physikalische Erdkunde* **4**: 241–330.

Iversen, J. (1936). *Biologische Pflanzentypen als Hilfsmittel in der Vegetationsforschung.* Mitt. Skallinglab. 4. Copenhagen.

Iversen, J. (1944). *Viscum, Hedera* and *Ilex* as climate indicators. *Geol. För. Stockholm Förh.* **66**(3): 463–83.

Iversen, J. (1949). The influence of prehistoric man on vegetation. *Danmarks Geologiske Undersøgelse Række IV*: 3, 6.

Iversen, J. (1954). The late-glacial flora of Denmark and its relation to climate and soil. *Danmarks Geologiske Undersøgelse Række II,* **80**: 87–119.

Iversen, J. (1956). Forest clearance in the Stone Age. *Scientific American* 194: 36–41.

Iversen, J. (1958). The bearing of glacial and interglacial epochs on the formation and extinction of plants. In Hedberg, O. (ed.) Systematics of to-day. *Uppsala Universitets Årsskrift 1958*, **6**: 210–215.

Iversen, J. (1973). The development of Denmark's nature since the last glacial. *Danmarks Geologiske Undersøgelse V, 7-C*, 1–126.

Jaccard, H. (1895). Catalogue de la Flore Vallaisanne. *Nouveaux Mem. Soc. Helv. Sci. Nat.* **34**: 1–472.

Jalas, J. & Suominen, J. (1972–94). *Atlas Florae Europaeae 1–10.* Helsinki, 2433 maps.

Jenny, H. (1941). *Factors of soil formation.* New York.

Jessen, K. & Lind, J. (1923). Det danske markukrudts historie. *Kgl. Danske Vidensk. Selsk. Skr. Naturv. Mathem. Afd. 8. Række 8*, 496 pp.

Jessen, K., Andersen, S. & Farrington, A. (1959). The Interglacial deposit near Gort, Co. Galway, Ireland. *Proceedings of the Royal Irish Academy B* **60**(3): 1–77.

Jonsell, B. (1988). Mikroendemism i det baltiska landhöjningsområdet. *Blyttia* **46** 65–73.

Jonsell, B. (1990). Fjällendemism och annan endemism i Skandinaviens flora. *Blyttia* **48**: 79–81.

Jørgensen, P. M. (1978). The lichen family Pannariacae in Europe. *Opera Botanica* **45**: 1–123.

Jørgensen, P. M. (1990). Trønderlav (*Erioderma pedicellatum*) – Norges mest gåtefulle plante. *Blyttia* **48**: 119–24.

Karlsson, T. (1986). The evolutionary situation of *Euphrasia* in Sweden. *Acta Universitet Upsala Symbolae Botanicae Upsaliensis* **27**(2): 61–71.

Kärnefelt, I. (1979). The brown fruticose species of *Cetraria. Opera Botanica* **46**, 1–150.

Kimura, M., Yokoi, Y. & Hogetsu, K. (1978). Quantitative relationships between growth and respiration II. Evaluation of constructive and maintenance respiration in growing *Helianthus tuberosus* leaves. *Botanical Magazine (Tokyo)* **91**: 43–56.

Kinloch, B. H., Westfall, R. D. & Forrest, G. I. (1986). Caledonian Scots pine: origins and genetic structure. *New Phytologist* **194**: 703–829.

Kinzel, H. (1982). *Pflanzenökologie und Mineralstoffwechsel.* Ulmer Verlag, Stuttgart.

Kjelvik, S. (1976). Varmeveksling og varmeresistens for noen planter, vesentlig fra Hardangervidda. *Blyttia* **34**: 211–26.

Klebelsberg, R. von (1913). Das Vordringen der Hochgebirgsvegetation in den Tiroler Alpen. *Österreichische Botanische Zeitschrift* **63**: 177–86, 241–52.

Kleiven, M. (1959). Studies in the xerophile vegetation in northern Gudbrandsdalen, Norway. *Nytt Magasin Bot.* **7**: 1–60.

Knaben, G. (1954). *Saxifraga osloensis* n.sp. a tetraploid species of the *Tridactylites* section. *Nytt Magasin Bot.* **3**: 117–38.

Knaben, G. (1959). On the evolution of the *Radicatum*-group of the *Scapiflora* papavers as studied in 70 and 56 chromosome species. *Opera Botanica* **2**(3): 1–74.

Knaben, G. (1979). Additional experimental studies in the *Papaver radicatum* group. *Botaniske Notiser* **132**: 483–90.

Knaben, G. (1982). Om arts- og rasedannelse i Europa under kvartærtiden. I. Endemiske arter i Nord-Atlanteren. *Blyttia* **40**: 229–35.

Köppen, W. (1920). Verhältnisse der Baumgrenze zur Lufttemperatur. *Met. Zeitschr.* **37**: 39–44.

Kolstrup, E. (1980). Climate and stratigraphy in northwestern Europe between 30 000 B.P. and 13 000 B.P., with special reference to the Netherlands. *Medd. Rijks. Geol. Dienst.* **32–15**: 181–238.

Kornas, J. (1987). Plant invasions in Central Europe: historical and ecological aspects. In Di Castri, F., Hansen, A. J. & Debussche, M. (eds.): *Biological invasions in Europe and the Mediterranean Basin*, pp. 19–36. Kluwer, Dordrecht.

Kornek, D. & Sukopp, H. (1988). Rote Liste in der Bundesrepublik Deutschland ausgestorbenen, verschollenen und gefährdeten Farn- und Blutenpflanzen und ihre Auswertung für den Arten- und Biotopschutz. *Schriftenreihe für Vegetationskunde. 19*, Bundesforschungsanstalt für Naturschutz und Landschaftsökologie. Bonn-Bad Godesberg.

Kotilainen, M. J. (1933). Zur Frage der Verbreitung des atlantischen Florenelement Fennoscandias. *Ann. Soc. Zool.- bot. Fenn. Vanamo 4, 1*: 1–75.

Kovarik, I. (1990). Some responses on flora and vegetation to urbanization in Central Europe. In Sukopp, H. & Hejny, S. (eds.): *Urban ecology. Plants and communities in urban environments*, pp. 45–74. SB Academic Publishing, The Hague.

Krahulec, F., Rosen, E. & van der Maarel, E. (1986). Preliminary classification and ecology of dry grassland communities on Ölands Stora Alvar (Sweden). *Nordic Journal of Botany* **6**: 797–809.

Krog, H. (1968). The macrolichens of Alaska. *Norsk Polarinst. Skr. 144*: 1–180.

Kullman, L. (1988). Holocene history of the forest-alpine tundra ecotone in the Scandes Mountains (central Sweden). *New Phytologist* **108**: 101–10.

Kutzbach, J. E. & Guetter, P. J. (1986). The influence of changing orbital parametres and surface boundary conditions on climatic simulations for the past 18 000 years. *Journal of Atmospheric Science* **43**: 1726–59.

Lagerberg, T., Holmboe, J. & Nordhagen, R. (1950–8). *Norges Planter, Vols I–VI2*. Tanum, Oslo.

Lambers, H. & Rychter, A. M. (1989). The biochemical background in variation of respiration rates. In Lambers *et al.* (eds.): *Causes and consequences of variation in growth rate and productivity of higher plants*, pp. 199–225. SB Academic Publishing, The Hague.

Landolt, E. (1977). Ökologische Zeigerwerte zur Schweizer Flora. *Veröff. Geobot. Inst. Rübel* **64**: 1–207.

REFERENCES

Lange, O. L. (1959). Untersuchungen über Warmehaushalt und Hitzeresistenz mauritanische Wüsten- und Savannenpflanzen. *Flora* **147**: 595–651.

Lange, O. L. & Lange, R. (1963). Untersuchungen über Blattemperaturen, Transpiration und Hitzeresistenz an Pflanzen mediterraner Standorte (Costa Brava, Spanien). *Flora* **153**: 387–425.

Larcher, W. (1984). *Ökologie der Pflanzen.* 4. Aufl. Ulmer Verlag, Stuttgart.

Larsen, E., Klakegg, O. & Longva, O. (1988). Brattvåg og Ona. Kvartærgeologisk kystsonekart 1220 III og 1220 IV-M. 1–50 000. *Norges Geol. Unders. Skr.* **85**: 1–41.

LaSalle, P., De Kimpe, C. R. & Laverdiere, M. R. (1985). Sub-till saprolites in southeastern Quebec and adjacent New England: erosional, stratigraphic and climatic significance. *Geological Society of America Special Paper* 197: 13–20.

Lid, J. (1985). *Norsk, svensk, finsk flora,* 5th edn. Det norske samlaget, Oslo.

Liden, M. (1991). Notes on *Corydalis* sect. *Corydalis* in the Baltic area. *Nordic Journal of Botany* **11**: 129–33.

Lindquist, B. (1932). Den vildväxande skogsalmens raser och deras utbredning i Nordvesteuropa. *Acta Phytogeographica Suecica* **4**: 1–46.

Lindroth, C. (1957). *The faunal connections between Europe and North America.* John Wiley & Sons, New York.

Linkola, K. (1916). Studien über den Einfluss der Kultur auf die Flora in den Gegenden nördlich von der Ladogasee. I. *Acta Societas Flora et Fauna Fennica* **45**(1): 1–432.

Linkola, K. (1921). Studien über den Einfluss der Kultur auf die Flora in den Gegenden nördlich von der Ladogasee. II. *Acta Societas Flora et Fauna Fennica* **45**(2): 1–490.

Longva, O., Larsen, E. & Mangerud, J. (1983). Beskrivelse til kvartærgeologisk kart 1019 II – M 1:50 000. *Norges Geol. Unders. Skr.* **48**: 1–66.

Lye, K. A. (1970). The horizontal and vertical distribution of oceanic plants in South West Norway. *Nytt Magasin Bot.* **17**: 25–48.

Lynge, B. (1938). Lichens from the West and North coasts of Spitsbergen and North-East Land. I. The macrolichens. *Skr. Norske Vidensk.- Akad. Oslo 1938, 6,* 136 pp.

Løve, A. (1961). *Hylandra* – a new genus of Cruciferae. *Svensk Botanisk Tidskrifter* **55**: 211–17.

Løve, A. & Løve, D. (eds.) (1963). *The North Atlantic biota and their history.* Pergamon Press, London.

McCarroll, D., Ballantyne, C. K., Nesje, A. & Dahl, S. O. (1995). Nunataks of the last ice sheet in northwest Scotland. *Boreas* **24**: 305–23.

McCree, K. J. (1970). An equation for the rate of respiration of white clover plants grown under controlled conditions. In Setlik, I. (ed.): *Predictions of measurements of photosynthetic productivity.* Proc. IBP/PP Techn. Meeting, Trebon 1969, pp. 221–30.

McVean, D. N. & Ratcliffe, D. A. (1962). *Plant communities of the Scottish Highlands; a study of Scottish mountain, moorland and forest vegetation.* Monographs of the Nature Conservancy, no. 1. Her Majesty's Stationery Office, London.

Manabe, S. & Hahn, D. G. (1977). Simulation of the tropical climate of an Ice Age. *Journal of Geophysical Research* **82**: 3899–911.

Matthews, J. R. (1955). *Origin and distribution of the British Flora.* Hutchinson, London.

May, R. M. (1974). *Stability and Complexity in Model Ecosystems.* 2nd edn. Princeton University Press, Princeton, NJ.

Mayr, E. (1909). *Waldbau auf naturgesetzliche Grundlage.* Parey, Berlin.

Mellor, R. & Wilson, M. J. (1989). Origin and significance of gibbsitic montane soils in Scotland, UK. *Arctic and Alpine Research* **21**: 417–24.

Mennema, J., Quene-Boterenbroot, A. J. & Plate, C. I. (1985). *Atlas van de Niederlandse Flora. 2. Zeltzame en vrij zeltsame plante.* Bohn, Scheltema & Holkema, Utrecht.

Meusel, H., Jäger, E. & Weinert, F. (1965). *Vergleichende Chorologie der zentraleuropäischen Flora.* Karten. Bd. I. G. Fischer Verlag, Jena.

Meusel, H., Jäger, E., Rauschert, S. & Weinert, F. (1978). *Vergleichende Chorologie der zentraleuropäischen Flora.* Karten. Bd II: 255–421. G. Fischer Verlag, Jena.

Miller, G. H. (1982). Quaternary depositional episodes, Western Spitsbergen, Norway: aminostratigraphy and glacial history. *Arctic and Alpine Research* **14**: 321–40.

Minyaev, N. A. (1968). Siberian taiga elements in the nortwest part of the European flora of the USSR. In Tolmatchev, A. I. (ed.): *Distribution of the flora of the USSR,* pp. 44–83. Israel Program for Scientific Translations, Jerusalem.

Miroslavov, E. A. & Kravkina, I. M. (1991). Comparative analysis of chloroplasts and mitochondria in leaf chlorenchyma from mountain plants grown at different altitudes. *Annals of Botany* **68**: 195–200.

Mitchell, F. J. G. (1993). The biogeographical implications of the distribution and history of the strawberry tree, *Arbutus unedo,* in Ireland. In Costello, M. J. & Kelly, K. S. (eds.): *Biogeography of Ireland: past, present, and future,* pp. 35–44. Occasional Publication of the Irish Biogeographical Society 2, Dublin.

Mitchell, G. F. & Watts, W. A. (1970). The distribution of Ericaceae in Ireland during the Quaternary epoch. In Walker, D. & West, R. G. (eds.): *Studies in the vegetational history of the British Isles,* pp. 13–21. Cambridge University Press, Cambridge.

Moe, D. & Odland, A. (1992). The influence of temperature climate on the vertical distribution of *Alnus incana* (Betulacae) in the Holocene in Norway. *Acta Botanica Fennica* **144**: 35–49.

Molau, U. (1991). The genus *Bartsia* (Scrophulariacae – Rhinanthoidea). *Opera Botanica* **102**: 1–99.

Mork, E. (1933). Temperaturen som foryngelsesfaktor i nord-trønderske barskoger. *Medd. Det Norske Skogforsøksvesen* **5**: 1–156.

Mork, E. (1938). Gran- og furufrøets spiring ved forskjellige temperatur og fuktighet. *Medd. Det Norske Skogforsøksvesen* **21**: 227–49.

Mucina, L., Krippelova, M., Zahlcherova, M. & Klotz, S. (1984). Proceedings of the 4th Symposium on Synanthropic Flora and Vegetation. *Acta Bot. Acad. Sci. Slov. Ser. A. Taxon. Geobot. Suppl. 1*, 362 pp. Bratislava.

Müller, K. (1954–7). *Die Lebermoose Europas.* Rabenhorst Kryptogamenflora von Deutschland, Österreich und der Schweiz, IV,1: 1–756, VI,2: 757–1365. Akad. Verl. Leipzig & Johnsen Reprint Corporation New York, London.

Myklestad, Å. (1993). The distribution of *Salix* species in Fennoscandia – a numerical analysis. *Ecography* **16**: 329–44.

Myklestad, Å. & Birks, H. J. B. (1993). A numerical analysis of the distribution patterns of *Salix* L. species in Europe. *Journal of Biogeography* **20**: 1–32.

Nathorst, A. G. (1892). Über den gegenwärtigen Stand unser Kenntnis der Verbreitung fossiler Glazialpflanzen. *Bihang Kungl. Svenska Vetensk.-Akad. Handl. 17. 3, 5*: 1–35.

Nesje, A., Anda, E., Rye, N., Lien, R., Hole, P. A. & Blikra, L. H. (1987). The vertical extent of the Late Weichselian ice sheet in the Nordfjord–Møre area, western Norway. *Norsk Geol. Tidsskr.* **67**: 125–41.

Nesje, A., Dahl, S. O., Anda, E. & Rye, N. (1988). Block fields in southern Norway: significance for the late Weichselian ice sheet. *Norsk Geol. Tidsskr.* **68**: 149–69.

Nesje, A., McCarroll, D. & Dahl, S. O. (1994). Degree of rock surface weathering as an indication of ice-sheet thickness along an east-west transect across southern Norway. *Journal of Quaternary Science* **9**:337–47.

Nilsson, T. (1972). *Pleictocen. Den geologiska och biologiska utvecklingen under istidsåldern.* Esselte Studium, Lund.

Noirfalaise, A. (1987). *Map of the natural vegetation of the member countries of the European Communities and the Council of Europe. Scale 1:3 000 000.* Luxembourg, 80 pp., 4 maps.

Nordal, I. (1985a). Overvintringsteori og evolusjonshastighet. *Blyttia* **43**: 33–41.

Nordal, I. (1985b). Overvintringsteori og det vestarktiske element i Skandinavisk flora. *Blyttia* **43**: 185–93.

Nordal, I. (1987). *Tabula rasa* after all? Botanical evidence for ice-free refugia in Scandinavia reviewed. *Journal of Biogeography* **14**: 377–88.

Nordhagen, R. (1931). Studien über die skandinavischen Rassen des *Papaver radicatum* Rottb. sowie einige mit denselben verwechselte neue Arten. *Bergens Mus. Årbok 1931. Naturvitensk. Rekke* **2**:1–50.

Nordhagen, R. (1935). Om *Arenaria humifusa* og dens betydning for utforskningen av Skandinavias eldste floraelement. *Bergens Mus. Årbok 1935. Naturv. Rekke*: 1–185.

Nyholm, E. (1954–69). *Illustrated moss flora of Fennoscandia.* Gleerup, Lund.

Nyholm, E. (1987–9). *Illustrated flora of Nordic mosses*, fasc. 1–2. Nordic Bryological Society, Copenhagen and Lund.

Oberdorfer, E. (1978). *Süddeutsche Pflanzengesellschaften. II.* G. Fischer Verlag, Stuttgart.

Oberdorfer, E. (1983). *Süddeutsche Pflanzengesellschaften. III.* G. Fischer Verlag, Stuttgart.

Odasz, A. M. (1991). Distribution and ecology of herbaceous *Pedicularis dasyantha* (Scrophulariaceae) in Spitsbergen, Svalbard Archipelago: Relation to present and past environments. *Striae* **34**: 145–52.

Oldfield, F. (1964). Late-Quaternary deposits at le Moura, Biarritz, Southwest France. *New Phytologist* **63**: 374–98.

Olsen, S. R. & Gauslaa, Y. (1991). Långskägg, *Usnea longissima*, hotad även i södra Norge. *Svensk Botanisk Tidskrift* **85**: 342–6.

Pearsall, W. H. (1950). *Mountains and moorlands.* New Naturalist Series. Collins, London.

Pedersen, B. (1990. Distributional patterns of vascular plants in Fennoscandia: a numerical approach. *Nordic Journal of Botany* **10**: 163–89.

Peglar, S. M. (1993). The development of the cultural landscape around Diss Mere, Norfolk, UK, during the past 7000 years. *Review of Palaeobotany and Palynology* **76**: 1–47.

Perring, F. H. (1967). The Irish problem. In Tüxen, R. (ed.): *Pflanzensoziologie und Palynologie*, pp. 257–68. Dr W. Junk, the Hague.

Perring, F. H. & Walters, S. M. (eds.) (1962). *Atlas of the British Flora.* Botanical Society of the British Isles, London. 3rd. edn. 1982.

Peterson, G. M., Webb, T., Kutzbach, J. E., van der Hammen, T., Wijmstra, T. A., & Street, F. A. (1979). The continental record of environmental conditions at 18 000 BP: an initial evaluation. *Quaternary Research* **12**: 47–82.

Pielou, E. C. (1991). *After the Ice Age. The return of life to glaciated North America.* University of Chicago Press, Chicago.

Pigott, C. D. (1970). The response of plants to climate and climatic change. In Perring, F. H. (ed.): *The flora of a changing Britain* pp. 32–44. Classey, Hampton, UK.

Pigott, C. D. (1975). Experimental studies on the influence of climate on the geographical distribution of plants. *Weather, May 1975*: 82–90.

Pigott, C. D. (1978). Climate and vegetation. In Clapham, A. R. (ed.): *Upper Teesdale. The area and its natural history*, pp. 102–21. Collins, London.

Pigott, C. D. (1981). Nature of seed sterility and natural regeneration of *Tilia cordata* near its northern limit in Finland. *Annales Botanici Fennici* **18**: 255–63.

Pigott, C. D. (1991). *Tilia cordata* Miller. Biological flora of the British Isles. *Journal of Ecology* **79**: 1147–1207.

Pigott, C. D. & Huntley, J. P. (1981). Factors controlling the distribution of *Tilia cordata* at the northern limits of its geographical range. 3. Causes of seed sterility. *New Phytologist* **87**: 817–39.

Plesnik, P. (1971). Die obere Waldgrenze in der Hohen und Belauer Tatra. *Vydavtelestvo Slovenskej Akad. Ved. Bratislava*, 202 pp.

Printz, H. (1933). Granens og furuens fysiologi og geografiske utbredelse. *Nyt Mag. Naturv.* **73**: 167–219.

Proctor, M. C. F. (1967). The distribution of British liverworts: a statistical analysis. *Journal of Ecology* **55**: 119–35.

Proctor, M. C. F. (1981). Physiological ecology of bryophytes. *Advances in Bryology* **1**: 79–166.

Pugsley, H. W. (1936). Enumeration of the species of *Euphrasia* L. Sect. *Semicalcaratae* Benth. *Journal of Botany (London)* **74**: 273–88.

Purvis, O. W. & Coppins, B. J. (1992). *Pilophorus* Th. Fr. (1857). In Purvis, O. W., Coppins, B. J., Hawksworth, D. L., James, P. W. & Moore, D. M. (eds.): *The lichen flora of Great Britain and Ireland*. Natural History Museum Publications/The British Lichen Society London.

Påhlsson, L. (1984). *Vegetasjonstyper i Norden*. Nordiska ministerrådet, 593 pp.

Quervain, A. de (1903). Die Hebung der atmosphärischen Isothermen in den Schweizeralpen und ihre Beziehungen zu den Höhengrenzen. *Gerlands Beitr. Geophysik* **6**: 481–533.

Ratcliffe, D. A. (1968). An ecological account of atlantic bryophytes in the British Isles. *New Phytologist* **67**: 365–439.

Ratcliffe, D. A. (ed.) (1977). *A nature conservation review. The selection of biological sites of national importance to nature conservation in Britain. I–II*. Cambridge University Press, Cambridge.

Ratcliffe, D. A. (1984). Post-Medieval and recent changes in British vegetation, the culmination of human influence. *New Phytologist* **98**: 73–100.

Raven, P. H. (1963). Amphitropical relationships in the floras of North and South America. *Quarterly Review of Biology* **38**: 151–77.

Reisigl, H. & Pitschmann, H. (1958). Obere Grenzen von Flora und Vegetation in der Nivalstufe der zentralen Ötztaler Alpen (Tirol). *Vegetatio* **8**: 93–129.

Rengel, Z. (1992). Role of calcium in aluminium toxicity. *New Phytologist* **121**: 499–513.

Roaldset, E., Pettersen, E., Longva, O. & Mangerud, J. (1982). Remnants of preglacial weathering in western Norway. *Norsk Geol. Tidsskr.* **62**: 169–78.

Robak, H. (1960). Spontaneous and planted forests in West Norway. In Sømme, A. (ed.): *Vestlandet. Geographical Studies*, pp. 17–34. Skr. Norges Handelshøyskole. Geografiske avhandlinger. 7.

Robinson, S. A., Yakir, D., Ribas-Carbo, M., Giles, L., Osmond, C. B. & Siedow, J. N. (1992). Measurements of the engagement of cyanide-resistant respiration in the Crassulacean acid metabolism plant *Kalanchoë daigremontana* with the use of on-line oxygen isotope discrimination. *Plant Physiology* **100**: 1087–91.

Rokoengen, K. & Rønningsland, K. M. (1983). Shallow bedrock geology and Quaternary thickness in the Norwegian sector of the North Sea between 60° 30′ N and 62° N. *Norsk Geol. Tidsskr.* **63**: 83–102.

Ruddiman, W. F. & Raymo, M. E. (1988). Northern Hemisphere climate régimes during the past 3 Ma: possible tectonic connections. *Philosophical Transactions of the Royal Society London B* **318**: 411–29.

Runge, M. & Rode, M. W. (1991). Effects of soil acidity on plant associations. In Ulrich, B. & Sumner, M. E. (eds.): *Soil Acidity*, pp. 183–202. Springer-Verlag, Berlin.

Rybníček, K. (1973). A comparison of present and past mire communities of Central Europe. In Birks, H. J. B. & West, R. G. (eds.): *Quaternary Plant Ecology*, pp. 237–61. Blackwell Scientific Publications, Oxford.

Rønning, O. I. (1964). *Svalbards flora*. Universitetsforlaget, Oslo.

Sakai, A. & Larcher, W. (1987). Frost survival of plants. *Ecological Studies* **62**: 1–321.

Salisbury, E. J. (1926). The geographical distribution of plants in relation to climatic factors. *Geographical Journal* **57**: 312–35.

Salisbury, E. J. (1935). Are most of the present British plants post-glacial immigrants from extra-British regions with possibly some human introductions? *Proceedings of the Royal Society of London* **118** :222–5.

Salisbury, E. J. (1961). *Weeds and aliens.* New Naturalist Series 43. Collins, London.

Salvesen, P. (1988). Sammenliknende dyrkingsforsøk med sørvestkyst-skyende planter. Del 1. Frilandsforsøk. *Blyttia* **46**: 145–53.

Salvesen, P. (1989). Sammenliknende dyrkingsforsøk med sørvestkyst-skyende planter. Del 2. Forsøk i kontrollert klima. *Blyttia* **47**: 143–53.

Sandgren, R. (1943). *Hippophae rhamnoides* L. i Sverige under senkvartar tid. *Svensk Botanisk Tidskrift* **37**: 1.

Sandmo, J. K. (1960). Problemer omkring furuskogens innvandring og senere tilbakegang. *Tidsskr. f. Skogbruk* **68**: 204–7.

Savile, B. D. O. (1972). Arctic adaptions in plants. *Canadian Department of Agriculture Research Branch Monograph* **6**: 1–81.

Savile, B. D. O. (1981). A naturalist looks at arctic adaptations. In Schudder, G. C. E. & Reveal, J. L. (eds.): *Evolution today. Proceedings of the second international congress of systematic and evolutionary Biology*, pp. 47–53.

Schauer, T. (1965-6). Ozeanische Flechten im Nordalpenraum. *Portugaliae Acta Biol.* **8**(8): 17–226.

Schofield, W. B. (1970). Phytogeography of northwestern North America: bryophytes and vascular plants. *Madroño* **20**: 155–207.

Schofield, W. B. (1974). Bipolar disjunctive mosses in the Southern Hemisphere, with particular reference to New Zealand. *Journal of the Hattori Botanical Lababoratory* **38**: 13–32.

Schuster, R. M. (1966-80). *The Hepaticae and Anthocerotae of North America east of the hundredth meridian*, I (1966), II (1969), III (1974), IV (1980). Columbia University Press, New York.

Scott, G. A. M. (1988. Australasian bryogeography: fact, fallacy and fantasy. *Botanical Journal of the Linnean Society* **98**: 203–10.

Segerstråle, S. G. (1957). On the immigration of glacial relicts of northern Europe, with remarks on their prehistory. *Soc. Scient. Fenn. Commentationes Biol. 16,* **16**: 1–117.

Sernander, R. (1896). Några ord med anledning av Gunnar Andersson: Svenske växtvärldens historia. *Botaniser Notiser 1896*: 114–28.

Sissons, J. B. (1974). The Quaternary of Scotland: a review. *Scottish Journal of Geology* **10**: 311–37.

Skre, O. (1971). Frequency distributions of monthly air temperature and their geographical and seasonal variations in Northern Europe. *Meld. Norges Landbrukshøgskole* **50**(9): 1–54.

Skre, O. (1972). High temperature demands for growth and development in Norway spruce (*Picea abies* (L.) Karst.) in Scandinavia. *Meld. Norges Landbrukshøgskole* **51**(7): 1–29.

Skre, O. (1979a). The regional distribution of vascular plants in Scandinavia with requirements for high summer temperatures. *Norwegian Journal of Botany* **26**: 295–318.

Skre, O. (1979b). The accumulated annual respiration as a measure of energy demands for growth in some Scandinavian plants. *Use of ecological variables in environmental monitoring. The National Swedish Environment Board, Report PM 1151*: 292–8.

Skre, O. (1983). Respirasjon, vekst og frømodning som skoggrensedannende faktor og basis for regional inndeling i Skandinavia. *Rapport Bot. ser. 1983–7*: 33–59. Universitet Trondheim, Vitenskapsmuseet, Trondheim.

Skre, O. (1993). Growth processes in mountain birch (*Betula pubescens* Ehrh. ssp. *tortuosa*) and other deciduous tree species, with particular reference to dark respiration as a factor limiting growth at low temperatures. Ph.D. thesis, University of Bergen.

Smith, A. J. E. (1978). *The moss flora of Britain and Ireland.* Cambridge University Press, Cambridge.

Smith, A. J. E. (1990). *The liverworts of Britain and Ireland.* Cambridge University Press, Cambridge.

Smith, G. L. (1972). Continental drift and the distribution of Polytrichaeceae. *Journal of the Hattori Botanical Laboratory* **35**, 41–9.

Smith, H. (1920). Vegetationen och dess utvecklingshistoria i det centralsvenska högfjällsområdet. *Norrlandsk handbibliotek* **9**, 1–238.

Soffer, O. (1990). The Russian Plain at the last glacial maximum. In Soffer, O. & Gamble, C. (eds.): *The World at 18 000 BP. High Latitudes*, pp. 228–52. Unwin Hyman, London.

Sollid, J. L., Carlson, A. B. & Torp, B. (1980). Trollheimen–Sundalsfjella, Oppdal Quaternary map 1:100 000. *Norsk Geogr. Tidsskr.* **34**: 177–89.

Stace, C. A. (1991). *New flora of the British Isles.* Cambridge University Press, Cambridge.

Stace, C. A. (ed.) (1975). *Hybridization in the flora of the British Isles.* Academic Press, London.

Steere, W. C. (1965). The boreal bryophyte flora as affected by the Quaternary glaciations. In Wright, H. E. & Frey, D. G. (eds.): *The Quaternary of the United States,* pp. 485–95. Princeton, NJ.

Steindorsson, S. (1963). Ice age refugia in Iceland as indicated by the present distribution of plant species. In Løve, A. & Løve, D. (eds.): *North Atlantic biota and their history,* pp. 303–20. Pergamon Press, Oxford.

Strauch, F. (1970). Die Thule-Landbrücke als Wanderweg und Faunenscheide zwischen Atlantik und Skandik im Tertiär. *Geologische Rundschau* **60**: 381–417.

Strauch, F. (1983). Geological history of the Iceland–Faeroe ridge and its influence on Pleistocene glaciations. In Bott, M. H. P. *et al.* (eds.): Structure and development of the Greenland–Scotland ridge, pp. 601–6. Plenum Press, London.

Strömberg, B. (1985). Revision of the lateglacial Swedish varve chronology. *Boreas* **14**: 101–5.

Størmer, P. (1969). *Mosses with a western and southern distribution in Norway.* Universitetsforlaget, Oslo.

Størmer, P. (1984). An eastern element within the Norwegian moss flora. *Rev. Bryol. Lichenol.* **1984**: 135–41.

Sukopp, H. & Hejny, S. (eds.) (1990). *Urban ecology. Plants and communities in urban environments.* SB Academic Publishing, The Hague.

Sutherland, D. G. (1984). The Quaternary deposits and landform of Scotland and the neighbouring shelves: a review. *Quaternary Science Review* **3**: 157–254.

Sutton, O. G. (1953). *Micrometeorology.* McGraw-Hill, New York.

Svensson, R. & Wigren, M. (1986). A survey of the history, biology and preservation of some retreating synanthropic plants. *Symbolae Botanicae Upsaliensis* **25**(4): 1–74.

Szafer, W. (1932). The beech and the beech forest in Poland. In Rübel, E. (ed.): *Die Buchenwälder Europas,* pp. 168–81. Veröff. geebot. Inst. Rübel, Zurich 8.

Sæbø, S. (1970). The autecology of *Rubus chamaemorus* L. II. Nitrogen economy of *Rubus chamaemorus* in an ombrotrophic mire. *Meld. Norges Landbrukshøgskole* **49**(9): 1–37.

Sørensen, R. (1982). Preboreal-Boreal isavsmeltning i Sørøst-Norge. *Rapport 17. Department Geol., Agric. Univ. Norway. Ås,* 68 pp.

Tallis, J. H. (1991). *Plant community history. Long-term changes in plant distribution and diversity.* Chapman & Hall, London.

Tamm, C. O. (1991). Nitrogen in terrestrial ecosystems. *Ecological Studies* **81**: 1–115.

Teppner, H. & Klein, E. (1985). Karyologie und Fortpflanzungs-modus von *Nigritella* (Orchidaceae–Orchidae) incl. *N. archiducis-joannis* spec. nov. und zweier Neukombinationen. *Phyton* **25**: 147–76.

Thorsrud, J. (1964). Om klima og fruktdyrking. *Frukt og bær* **16**: 8–12.

Tranquillini, W. (1979). Physiological ecology of the alpine timberline. *Ecological Studies* **31**: 1–133.

Troll, K. (1925). Ozeanische Zuge im Pflanzenkleid Mitteleuropas. *Freie Wege vergleichender Erdkunde. Festgabe an Dryglaski.* Münich & Berlin.

Troll, C. (ed.) (1972). *Geoecology of the high-mountain regions of Eurasia* (Erdwissenschaftliche Forschung 4.). Franz Steiner Verlag, Wiesbaden.

Tuhkanen, S. (1980). Climatic parameters and indices in plant geography. *Acta Phytogeographica Suecica* **67**, 1–110.

Valentine, D. H. & Løve, A. (1958). Taxonomic and biosystematic categories. *Brittonia* **16**: 153–66.

van Rompaey, E. & Delvosalle, L. (1978). *Atlas de la flore Belge et Luxembourgoise. Commentaires.* Meise, 116 pp.

van Rompaey, E. & Delvosalle, L. (1979). *Atlas de la flore Belge et Luxembourgoise. Pteridophytes et Spermatophytes.* Meise, 1542 maps.

Vestergaard, P. & Hansen, K. (eds.) (1989). Distribution of vascular plants in Denmark. *Opera Botanica* **96**: 1–162.

Vorren, K.-D. (1979). Anthropogenic influence on the natural vegetation in coastal North Norway during the Holocene. Development of farming and pastures. *Norwegian Archaeological Review* **12**: 1–21.

Vorren, K.-D. (1986). The impact of early agriculture on the vegetation of northern Norway. A discussion of anthropogenic indicatores in biostratigraphical data. In Behre, K.-E. (ed.): *Anthropogenic indicators in pollen diagrams*, pp. 53–64. Balkema, Rotterdam.

Vorren, T. O., Vorren, K.-D., Alm, T., Gulliksen, S. & Løvlie, R. (1988). The last deglaciation (20 000–11 000 B.P.) on Andøya, northern Norway. *Boreas* **17**: 41–77.

Walters, S. M. (1978). British endemics. In Street, H. E. (ed.): *Essays in plant taxonomy*, pp. 263–74. Academic Press, London.

Wardle, P. (1974). Alpine timberlines. In Ives, J. D. & Barry, R. G. (eds.): *Arctic and Alpine environments*, pp. 371–402. Methuen, London.

Warming, E. (1888). Om Grønlands vegetation. *Medd. Grønland* **12**: 1–223.

Watson, A. Miller, G. R. & Green, F. H. W. (1966). Winter browning of heather (*Calluna vulgaris*) and other moorland plants. *Trans. Proceed. Bot. Soc. Edinburgh* **40**: 195–203.

Watts, W. A. (1988). Late-Tertiary and Pleistocene vegetation history – 20 My to 20 ky. Europe. In Huntley, B. & Webb, T. (eds.): *Vegetation history. Handbook of vegetation science, Vol. 7*, pp. 155–92. Kluwer, Dordrecht.

Webb, D. A. (1985). What are the criteria for presuming native status? *Watsonia* **15**: 231–6.

Welten, M. & Sutter, R. H. C. (1982). *Verbreitungsatlas der Farn- und Blutenpflanzen der Schweiz, Vols. 1 & 2.* Burkhauser Verlag, Basel.

Wendelberger, G. (1959). *Artemisia oelandica* (Besser) Kraschen – ein Waldsteppenrelikt auf Øland. *Botanische Jahrbuch* **78**: 253–334.

West, R. G. (1977). *Pleistocene geology and biology.* 2nd edn. Longman, London.

Willerding, U. (1986). Zur Geschichte der Unkrauter Mitteleuropas. *Göttinger Schriften Vor- und Frühgeschichte* **22**, 382 pp.

Wilson, E. O. (1991). Rain forest canopy: The high frontier. *National Geographic Magazine, December 1991:* 78–107.

Wilson, P. (1991). Europe's endangered arable weeds. *Shell Agriculture* **10**: 4–6.

Woodward, F. J. (1987). *Climate and plant distribution.* Cambridge University Press, Cambridge.

Woodward, F. J. & Pigott, C. D. (1976). The climatic control of the altitudinal distribution of *Sedum rosea* (L.) Scop. and *S. telephium* L. Field observations. *New Phytologist* **74**: 323–34.

Yeo, P. F. (1978). A taxonomic revision of *Euphrasia* in Europe. *Botanical Journal of the Linean Society* **77**: 223–334.

Zwinger, A. H. & Willard, B. E. (1972). *Land above the trees: A guide to American Alpine Tundra.* Harper & Row, New York.

Index

Page numbers in italics denote illustrations or tables.

270